Praise for Z

"The narrative is briskly conversational: We're on the porch, shooting the breeze with a knowledgeable neighbor. Mr. Berman's avowed goal in writing this book, he says, was 'to open a window onto the enormous universe of omnipresent energies.' Once that window is thrown open, it is hard to look at the world the same way." *Wall Street Journal*

"Nimbly busts common myths... Erudite but never stuffy, Berman writes with enthusiasm and clarity, making this an informative and digestible read for the science-curious." *Booklist*

"[Berman's] an unfailingly congenial explainer, always ready with the kinds of fascinating facts his readers might have missed in school." *Christian Science Monitor*

"Captivating... fear not the long-winded scientific discourse: Berman zings through historical and scientific adventures." *American Scholar*

"An enthusiastic account of all the light we cannot see from a science popularizer with a knack for presenting hard facts clearly and stylishly ... A guide for laymen written with gusto and assurance." *Kirkus*

"Astronomy writer Berman runs through a fascinating history of the rainbow's invisible bands in this breezy, accessible read... In the style of a favorite professor, Berman injects bits of odd humor and captivating tangents into this complex but familiar topic." *Publishers Weekly*

"[Berman] excels at making complex concepts accessible for lay readers ... this is a great option for those curious about history, theories, and function of everyday things." *Library Journal*

"Explaining light using anecdotal history and colloquial explication, *Zapped* makes entertaining sense out of what could be dry math and physics." *Shelf Awareness*

ALSO BY BOB BERMAN

Zoom

The Sun's Heartbeat

Biocentrism (with Robert Lanza, MD)

Shooting for the Moon

Strange Universe

Cosmic Adventure

Secrets of the Night Sky

ZAPPED

FROM INFRARED TO X-RAYS, THE CURIOUS HISTORY OF INVISIBLE LIGHT

BOB BERMAN

A Oneworld Book

First published in Great Britain, the Republic of Ireland and Australia
by Oneworld Publications, 2018

This edition published by arrangement with Little, Brown and Company, a division of Hachette Book Group, Inc., New York, New York, USA.
A CIP record for this title is available from the British Library

ISBN 978-1-78607-373-0
eISBN 978-1-78607-374-7

Printed and bound in Great Britain by Clays Ltd, St Ives plc

Oneworld Publications
10 Bloomsbury Street
London WC1B 3SR
England

Contents

It is obvious that we must attach a deeper and much wider meaning to the word light than has hitherto been ordinarily understood.

—EDITORIAL IN *THE LANCET*, FEBRUARY 22, 1896

ZAPPED

INTRODUCTION

It's everywhere.

At this moment, as you sit quietly reading this book, you are awash in it. At work, it's emanating from your electronic devices; step outside for lunch, and the sun bathes you in it. You may receive an extra dose of it when you visit your doctor, pass through security at the airport, or drive through city streets, but minuscule amounts of it are with you always. You cannot see, hear, smell, or feel it, but there is never a single second when it is not flying through your body. Too much of it will kill you, but without it you wouldn't live a year.

"Invisible light" seems like a contradiction. Like Simon and Garfunkel's "The Sound of Silence," it's an oxymoron. We think of light, by definition, as something seen, something that enables our seeing, illuminating the darkness. Unlike dogs, who sniff in order to "get" what an object is, we depend on vision above all our other senses. We rely on light to tell us about our surroundings.

But just as there are frequencies of sound audible to other animals that we cannot hear, there is a whole world of light outside our range of vision, a world that is humming with activity. Though we rarely think about this invisible world, our way of life depends on it. It's because of invisible light that you can do things such as send a text message, use GPS to find your way to a friend's house, listen to the radio, or microwave a frozen pizza. Invisible light shows us things we would never otherwise see, including our own skeletons and brains and the history of our universe.

I was reminded of just how much we rely on invisible light, and how mysterious it remains to us, during a visit from my sister and her family. It was a lazy summer afternoon, and we were sprawled across a few couches sharing a bowl of popcorn. My niece, her shoulders crimson after a day outdoors, was chatting on her cell phone, holding up a promising finger to her mother, who was scolding her for not using sunscreen. My brother-in-law, meanwhile, was asking my opinion on an article he'd read proposing that Wi-Fi be banned in schools because of its dangers. We were all depending on invisible light (for the microwave popcorn, the cell-phone service) while being concerned that it might harm us (sunburn, mysterious Wi-Fi health threats) and confused about what to do to protect ourselves.

We need it, and it surrounds us, yet we remain uneasy about living with invisible light, partly because we fear what is unknown. After all, most of us don't know much about "all the light we cannot see." This book aims to change that.

My hope is to expose the hidden side of the spectrum, to make the invisible (at least temporarily) visible and vivid to you. As you'll see, each of the different varieties of invisible light—from gamma rays to infrared to ultraviolet—has its own characteristics and peculiarities, as distinct as red is from blue. We'll meet rays that pass through solid matter instead of being reflected by it and others that cause water to boil. Some come from deep space and zoom through astronauts' brains; others are left over from the birth of the universe. You may be surprised to find how much of our world and our history is touched by invisible rays. They rescued lives when the *Titanic* went down. They help determine the daily weather. Some can produce sudden, lethally destructive damage in human bodies.

We'll split our exploration into two equally important parts. In some chapters we'll go back in time to meet the pioneering scientists who first "saw" the invisible. Until the eighteenth century, no one had the faintest idea that there might be such a thing as light that cannot be detected by human vision, and until the nineteenth century no one had any proof of its existence. But once the discoveries of invisible light began, they avalanched, until most of today's indispensable aspects of life, the ones we take for granted, depend on them.

In other chapters we'll explore how these phantoms affect our lives and our bodies as they provide their near-magical conveniences, from the cell phone in your pocket to the radio in your car. What role does invisible light play in our medicine, our technology, and our culture in the twenty-first century? What new opportunities for its use are on the horizon?

Like my family lazing around the living room that summer afternoon, you probably have questions about how invisible light is affecting your health. What are the microwaves from your cell phone doing to your brain? What is radiation, and how much of it are you exposed to? Which invisible ray causes the most annual deaths? Which foods are most radioactive? This book will answer all those questions, clarifying once and for all the controversial claims about radiation's health consequences. Some of the things you'll learn will soothe you (ultraviolet light can decrease your risk of cancer), and some will shock you (a single whole-body CT scan delivers more radiation than was received by Hiroshima survivors a mile from ground zero), but in every case context is key. Myths will be busted, and wild facts will abound.

CHAPTER 1

Light Fantastic

If God really did say, "Let there be light," it was no small housewarming present. There is a *lot* of light in the universe—one billion photons of light for every subatomic particle. In terms of individual items in the cosmos, including the components of atoms, photons constitute 99.9999999 percent of everything. The universe is literally made of light. And that includes not only ordinary everyday visible light but also the vast majority of light—the kind we *cannot* see.

Light is an astonishing entity, and the quest to understand it has obsessed the greatest thinkers in disparate cultures through the centuries. The ancient Greeks, probably by sheer dumb luck, were the first to hit upon a key aspect of visible light—that it does not exist independent of us as observers. Physics now tells us that light is made up of intertwined magnetic and electrical fields. Since neither magnetism nor electricity is visible to our eyes, *light is inherently invisible.*

When we look at a bright orange sunset, we're not directly perceiving actual light. Rather, the energy reaching us from those electromagnetic pulses stimulates billions of neurons in our retinas and brains, which then fire to arouse a complex neurological architecture that produces in us the *sensation* of orange.

An entire biological empire is thus as essential to the existence of brightness and colors as the photons themselves.

The Greeks didn't know anything about brain structure, of course, yet they still figured out that light is a sensation, with no existence independent of the observer—which was either amazingly perceptive or just a lucky guess. But the Greeks had light's direction wrong. Knowing that its speed appeared instantaneous, they didn't imagine that a pulse of light originating in a candle sped in our direction until it struck our eyes. On the contrary, they regarded light as a ray traveling outward from our pupils. This belief, that our eyes project an illuminating beam, was universally embraced for more than a millennium. Even so, a few early iconoclasts envisioned eyesight as an *interplay* between this supposed eye ray and something emitted by other sources.

The classical thinker who came closest to the truth about light was the Roman Lucretius, who in the first century BCE, in his *On the Nature of Things,* wrote, "The light and heat of the sun are composed of minute atoms which, when they are shoved off, lose no time in shooting right across the interspace of air."

Lucretius's view of light as particles—later supported by Isaac Newton—included that profound "lose no time" characterization, showing that he believed light moved immeasurably fast. But whether scientists considered its speed merely superquick or instantaneous, light remained popularly regarded as a phenomenon that originates in the eye for centuries to come.

The first true breakthrough came from the mathematician and astronomer Alhazen—formally known as Abu Ali al-Hasan ibn al-Hasan ibn al-Haytham—who lived in Egypt during the golden age of Arab science. Around the year 1020, when the rest of the world was in the intellectual coma of the Dark Ages, Alhazen said that vision results solely from light entering the eye; nothing emanates from the eye itself. His popular pinhole camera obscura, which drew astonished yelps of wonder when visitors observed the phenomenon, gave weight to his arguments, for here was a full-color "motion picture" of nature splayed out on the walls. But Alhazen went much further. Light, he said, consists of streams of tiny, straight-moving particles that come from the sun and are then reflected by various objects. Sounds simple, perhaps, but Alhazen's spot-on conclusions were six centuries ahead of anyone else's.

The Renaissance turned up the juice on the "What is light?" debate, which eventually took on the quality of a food fight. In the late seventeenth century, Newton joined astronomer Johannes Kepler in arguing that light is a stream of particles, while men such as Robert Hooke, Christiaan Huygens, and, soon, Leonhard Euler insisted that light is a wave. But what is it a wave *of?* They thought there had to be a substance doing the waving, so these Renaissance scientists decided that space was filled with a plenum (later called an ether), an invisible substance that facilitated the movement of magnetic and electrical energy.

One obvious fact managed to sway many in favor of Newton's particle idea. When light from the sun passes a sharp edge, such as the wall of a house, it casts a sharp-edged shadow on nearby objects. That's what particles moving in a straight line should do. If instead light were made of waves, it ought to spread out—*diffract*—as ocean waves do when passing a jetty. To the particle proponents, the existence of sharp-edged shadows, combined with Newton's reputation as a genius, made the wave proponents seem like nut jobs.

The particles-versus-waves controversy eventually took a curious turn. As if some wise King Solomon ruled nature, *everyone* was soon declared right. The biggest breakthrough came from Scottish physicist and mathematician James Clerk Maxwell, who in 1865 showed that all light is a self-sustaining wave of magnetism with an electric pulse wiggling at right angles to it. One type of pulse stimulates the other, so that both the electrical and the magnetic waves continue indefinitely. From then on, science called light an *electromagnetic* phenomenon.

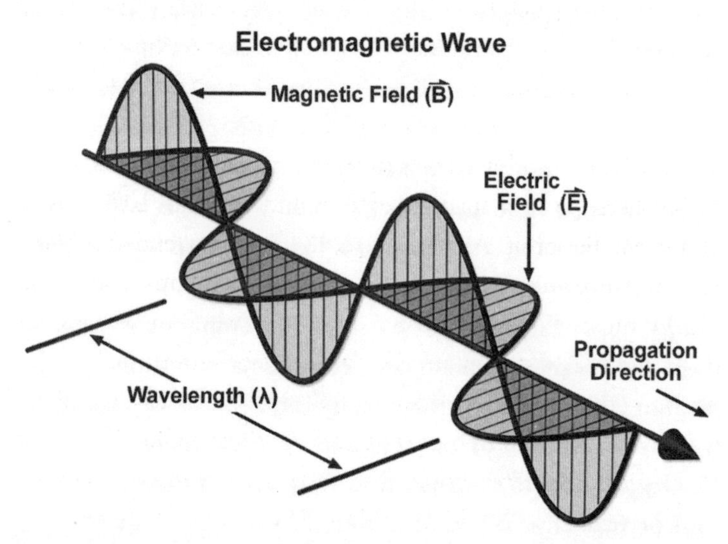

All light consists of a dual wave. A magnetic pulse is accompanied by an electric pulse positioned at a ninety-degree angle to it. *(Molecular Expressions at Florida State University)*

But where did this phenomenon come from? In 1896, the Dutch physicist Hendrik Lorentz figured out that the existence of a strange phenomenon—light splitting itself in two within a strong magnetic field—must mean that the rapid motion of some tiny, unknown, negatively charged particle has to be the source of all light everywhere in the universe. A year after he drew this astoundingly prescient conclusion, the first subatomic particle was discovered. This was the electron, *whose movements are indeed the sole creator of all light.* For predicting the existence of the electron before its actual discovery, Lorentz won the Nobel Prize in 1902.

How exactly is light—any light, all light—created? When an atom gets struck by energy—from a quick zap of electricity or a collision with a stray electron or the introduction of heat—its wiggling motion is jolted into a greater speed. This extra energy excites the atom's electrons, which give a figurative yelp and jump to an orbit farther from the nucleus. They don't like to be there, so in a fraction of a second they fall back into a closer, smaller orbit. As they do so, the atom surrenders a bit of energy. Since energy is never lost under any circumstances, this energy must manifest itself in some other way. And it does. A bit of light, a *photon,* materializes out of the emptiness as if by magic—then instantly rushes away at its famous breakneck speed. That's the only way light is ever born. *Out of seeming nothingness, whenever an electron moves closer to its atom's center.* Simple, really.

So light can be thought of as a set of two waves, one of electricity and one of magnetism, or as a weightless particle called a *photon*. Taking our cue from Albert Einstein, we might visualize a photon as a tiny bullet, an energetic speck with no mass, weighing nothing and with the curious property of being unable to ever stop moving. Nowadays, most people who think about such things (we science nerds) find it easiest to visualize light as a wave when it's en route from point A to point B and as a photon when it finishes its journey by crashing into something. But one may call light a photon or a wave and be equally correct.

The twentieth century brought us quantum theory, which—in addition to showing that solid objects such as electrons can behave as energy waves, too—revealed something extremely weird: when an observer uses an experimental apparatus to determine the location of photons or subatomic particles such as electrons, these entities always behave as particles and do things only particles can do, such as pass through one little hole or another but not both at once. But when no one's measuring where exactly each photon is situated, they behave as waves that blurrily pass through both holes in a barrier simultaneously to create an interference pattern on a detector located beyond the openings—which only waves can do.

Thus the observer and, weirdly, the information in his or her mind plays a critical role in whether light exists as a wave or as a tiny discrete object. The same is true for particles of matter. What you see depends on how you observe and what you know. Most physicists now think that a human consciousness is required to make a photon or an electron's "wave function" collapse so that it occupies a particular place as a particle. Otherwise it's just a theoretical object with neither location nor motion.

Just a century ago, the local realism mind-set of science, and even common sense, held that all objects, including atoms and photons, have an existence independent of our observation of them. But that's been replaced by a more modern view—that our observation itself is necessary for the very existence of photons and electrons, a spooky prospect.

But does an electron's wave function collapse and turn into an actual particle if a *cat* is watching? Would light always be waves and never discrete photons if no humans were around? Our best answers are "Who knows?" and "Yes" respectively, but obviously this whole business is Wonderland-strange.*

Let's make this strangeness clearer. A century ago, if we detected a bit of light (or even a physical particle) arriving at an instrument with which we could measure its incoming direction, we'd have confidently plotted out its previous path. No longer. Now we say that *it had no path* before we started to observe it. It possessed no real existence as an actual photon or electron or whatever it was. Rather, its observed existence is its only existence. Observation establishes reality. Nothing else is certain. As the late physicist John Wheeler put it, "No phenomenon is a real phenomenon until it's an observed phenomenon."

Which brings us to our next question: why can we observe some kinds of light with the naked eye and not others?

---

* My friend Matt Francis, an electron microscopist, is training his dog to recognize and respond to light displayed as a wave pattern on a screen as opposed to a series of particles. If he succeeds in teaching the dog to bark when observing waves and remain silent when observing only particles, he may be able to settle the matter and determine whether a *dog's* consciousness can "collapse" a photon into its particle configuration. Yes, such issues actually obsess some of us.

CHAPTER 2

Now You See It, Now You Don't

All forms of light, from the visible to the invisible, reside on the electromagnetic spectrum. Along this range there are many kinds (and colors) of light, and each variety can be distinguished from the others by two straightforward properties.

The first is *wavelength*. The length of each individual light wave varies from the tiniest fraction of an inch to more than a mile and spans everything in between.

The second is *frequency,* meaning the period of time it takes the wave to pass you and be replaced by the next wave, as if you were seated in a reviewing stand watching the light parade before your eyes.

Think of an ocean wave. In the open sea, a typical wave is around one hundred yards, or ninety-one meters, long—roughly the length of a football field. Its frequency is a bit less than one second. This means that each wave's peak requires nearly a second to pass any given point and be replaced by a trough, which in turn is followed by the peak of the next wave.

Science can identify any wave, or any particular type of light, by either its length or its frequency. For example, each wave of green light at a traffic signal has a length of 530 nm (or nanometers, meaning 530 billionths of a meter), which is about one millionth of an inch. These tiny waves have a frequency of 530 THz, or terahertz, which means that 530 trillion of them pass your eye each second. (That the number 530 appears in both wavelength and frequency is a coincidence; the matchup is true of green light but not of any other color).

When the signal turns red, you perceive waves of a longer length—twice as long, in fact. Each crimson-light wave is around two millionths of an inch from crest to crest. Red light has the longest waves of all visible light, but they're still smaller than most germs in our body. These waves vibrate more slowly than green light, too, with "just" 450 trillion wave pulses occurring per second. What's important is that all the light we see has wavelengths somewhere between 400 and 700 nm, which used to be expressed as 4,000–7,000 angstroms. *The light we* cannot *see has wavelengths either shorter or longer than that.*

Short waves pulse, or change, more quickly than long ones, and this gives them more power, or energy. As a result, while the light we *can* see is too weak to break atoms apart, fast-vibrating light such as ultraviolet light can indeed strip an atom of one or more of its electrons, which alters molecules and can lead to consequences such as carcinogenesis.

Invisible light has generally been named according to either its wave size or its position on the spectrum compared to the visible colors. Thus *infra*red light occupies a place just *before* the visible red light in the spectrum, meaning its waves are a bit longer than the red-light waves coming at you when you're stopped at a traffic light. By contrast, *ultra*violet light lies just *after* violet light, and its waves are slightly shorter.

The weakest kind of light is a radio wave. The longest radio waves measure a thousand miles from crest to crest. By contrast, the distance from one visible light wave to the next is just one millionth of a meter, or one hundred-thousandth of an inch. A few hundred *trillion* visible waves pass you every second. Even more mind-bogglingly short and fast are gamma rays, the strongest kind of light, with crests spaced just a trillionth of a meter apart and frequencies of a billion trillion per second. All other parts of the spectrum lie in between radio waves and gamma rays.

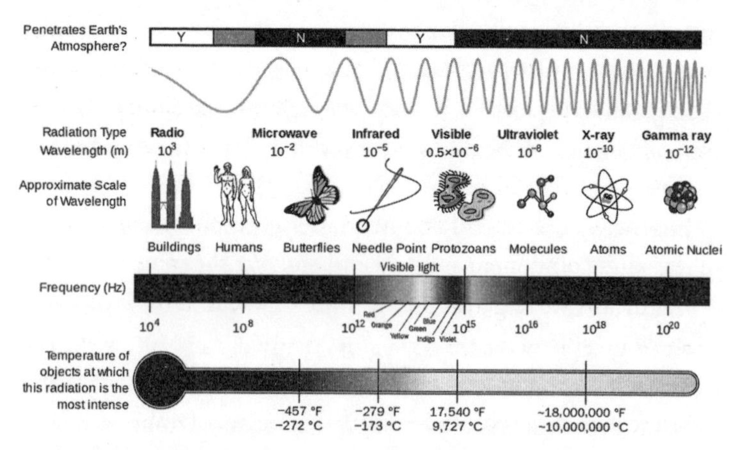

Visible light occupies only a tiny part of the electromagnetic spectrum. *(Wikimedia Commons)*

Except for the dim glow of the stars, all light is ultimately solar. Moonlight is reflected sunlight. The aurora borealis results from solar particles electrically stimulating the sparse oxygen atoms a hundred miles up. Candlelight and other kinds of flame require combustible materials such as coal, wood, and oil, which are forms of stored energy from long-dead plants and animals that would never have existed without the sun.

In our era we also create light using electricity, but that, too, comes from burning oil, gas, coal, or hydropower generated by falling water, which would never circulate back to higher altitudes without everyday solar warmth. Only nuclear power and starlight are independent of the sun, and stars emit exactly the same visible and invisible rays as our own sun does. Stars differ only in their proportions: hot, massive stars emit copious ultraviolet rays and blue light, whereas the more numerous lightweight stars give off copious reds, oranges, and infrared radiation, with very few UV rays. Rather poetically, our eyes *see only the colors the sun emits most strongly*. Our retinas are designed to perceive sunlight's most abundant energies and nothing else. So we really do have a sun bias. In a way, we scan the universe through the sun's eyes.

As we learned in science classes at school, the sun's white light is merely our retinal and neurological response to receiving all the sun's component spectral emissions at the same moment. White means we're getting it all. In a very real sense, white is a rainbow in a blender.

Indeed, if a scientist looks through a spectroscope, which reveals the true colors in the object she is studying by "unscrambling" them, a cloud that appears white to the unaided eye will resemble a vivid rainbow. The instrument reveals that a white cloud is actually composed of red, orange, yellow, green, blue, indigo, and violet light, and when those colors hit our eyes all at once, we see white. Studies conducted way back in the eighteenth century showed that even if a few of those colors are absent—orange and violet, say—we will still see white. Turns out all that's needed to make white are blue, red, and green combined in equal measure. These are called light's primary colors (totally different from yellow, cyan, and magenta, the primary colors of paint and pigment). So if we see white, it means we're receiving red, blue, and green light simultaneously.

If they're combined unequally, those same three primary colors will create others. Your computer and your TV use this trick all the time. If a friend sends you a digital photo of autumn foliage, you might see that some leaves appear deep reddish-purple on the screen. Your computer creates this effect by mixing, say, eighteen parts blue, seven parts red, and one part green. Just three hues, commonly called RGB, for red, green, and blue, can combine to create every possible color.

Some combinations are not logically obvious. Guess what mixture is needed to create yellow light? The answer is an equal blend of green and red. This surprises many people, because it seems logical that mixing red and green light would create a sort of reddish-green sensation. But no. Yellow is our eyes' reaction to red mixed equally with green.

On the other hand, if we mix red and green *paint* instead of red and green light, we won't get yellow but rather some muddy brownish-black hue. That's because entirely different rules apply to paint. Pigment does not glow on its own; we see a paint color only because some external white light (e.g., the ceiling lamp or the daylight streaming in from windows), is hitting the palette or painting, where the pigments absorb some of the white light's colors and reflect others. Yellow paint, for example, looks yellow because its chemical absorbs white light's blue component but reflects the light's red and green components—and the combination of red and green light always yields the sensation of yellow.

We see pigments and paints by a *subtractive process*. If you've ever painted, you may have had the frustrating experience of trying to create a color by mixing others—and ending up with a palette covered in brown. That's because each new color you add introduces further subtractions from the white light illuminating the room, reducing reflections from the canvas so that less light reaches your eyes. Further pigment combinations invariably darken the image. Add too many pigments, and the result is muddy brown or black, because by that point nearly all light is absorbed by the paint's molecules and nothing is reflected to our eyes. But light is a different story. Adding more light always brightens an image.

The experience of vision is a symbiotic event. This is so fundamentally important yet so little known that it bears repeating from the previous chapter: *by itself, light has no color or brightness.* Light is merely a wave of magnetism and, at right angles to it, a wave of electricity. So the *real* external world is as utterly invisible as radio waves. Without humans to perceive it, "external reality" is nothing but a complex jumble of various blank energy frequencies. But when stimulated by these invisible frequencies, our six million cone-shaped, photon-sensing retinal cells respond, each to a rather narrow set of predetermined vibrations. Thus stimulated, they send an electrical signal at 250 miles per hour up the optic nerve until several hundred billion neurons in the rear of the brain fire in a continuous, complex way. The result is an image perceived in the brain as a color, such as blue.

Bottom line: *the "external world" is an internal experience.* On their own, colors are not "out there." When no one is looking, a sunset has neither color nor brightness. It is an invisible mélange of electrical and magnetic pulses.

Some of us undergo an unusual subjective experience when confronted with the sun's photons. Deuteranopes—the 10 percent of males who lack the green retinal receptor—see far fewer colors than the rest of us do. To them, shades of red and green can look identical—to each other as well as to what we perceive as yellow, the combination of the two. Theirs is a world of blues and yellows. They don't understand why the rest of us are so enchanted by a rainbow, because to them it's merely a swath of two colors. These people can easily run traffic lights if the bulbs are in unfamiliar positions. Turns out dogs and elephants are deuteranopes, too. It's one of the reasons we should never let them drive.

In sunlight or bright artificial illumination, our best retinal mechanism—the cone-shaped retinal cells that deliver full-color vision three times sharper than our prized 1080p high-definition TV—is able to operate. This is why we see best when we're looking straight ahead: the densest concentration of cone cells lies in the center of the retina. The keenest vision also lies in the green part of the spectrum, right in the middle of our visual color range. Since green is our most readily perceived color as well as the one the sun emits most strongly, it perhaps deserves a few moments of our time.

Our eyes can distinguish between wavelengths that differ by just one nanometer, but only in the green section of the color palette. When it comes to reds and violets, our visual acumen is only one-tenth that keen. This means that human vision can detect around fifty different shades of green when they're positioned side by side.

The test for color sensitivity is usually performed using a split screen on which slightly different wavelengths are displayed next to each other. When there is less than one nanometer difference between them, the observer reports a unity of appearance. As slightly different tints are displayed on the halves, a critical point arrives when a separate color is sensed: the screen suddenly seems sharply divided in two.

Researchers test animals using the same setup, rewarding

dogs if they push their noses onto the part of the screen where the color is different (and presumably wiping all those nose prints clean at some point). Similar studies show that cats see colors, too. Whether in man, monkey, or Maltese, all this color perception is called photopic vision. Photopic vision is the full-color, full-sharpness vision we enjoy whenever there is ample light.

All this photopic vision from our bright-light neurological architecture has interesting idiosyncrasies. We already know that we see yellow when we receive an equal dose of green and red light. But this contrasts with our subjective experience of light's other primary colors. We do indeed perceive reddish blue as the sensation of purple, and we apprehend green plus blue as aquamarine, or cyan, which does resemble both of its components. Yet thanks to the quirks of human eye-brain architecture, we cannot perceive reddish green or yellowish blue.

Another oddity involves sensitivity. The daytime sky actually appears violet to a number of animal species, but our human retinas are so insensitive to the wavelengths at the fringe of the spectrum that we instead see the next most prevalent color emitted by the sky—blue. The animals and insects that can see the true violet color of the sky are also able to perceive ultraviolet light. Birds in particular readily see far more colors than we do. Their ability to perceive ultraviolet light allows them to detect the glow of mouse urine in a field far below.

Yet even with all the vagaries of our photopic, or full-color, vision architecture, everything *really* changes when the sun sets. When the photon count declines, our eyes shift from normal photopic vision to *scotopic* vision, or night vision.

Of all life's solar-induced rhythms, this daily plunge into darkness is the most familiar. Yet how much do you know about the vision you've used every night of your life? As darkness descends, our pupils expand to triple their previous size—as much as seven or eight millimeters in diameter if we're young. (Our pupils generally cannot expand beyond four or five millimeters after we pass the age of fifty.) Nineteenth-century astronomers, eager to be the first to detect new galaxies, or nebulae, would sometimes use belladonna to enlarge their pupils beyond their usual size in hopes of letting in more light and being able to see more through their telescopes.

At the same time, in dim light, photochemical changes in the retina greatly boost its sensitivity. Before our scotopic vision has fully switched on—that is, in dim but not truly dark conditions, such as in twilight or in a room lit only by a nightlight—photopic vision still operates, but not well. Only green objects keep their color, and the colors at the ends of the spectrum, red and violet, appear gray. We experience this in full moonlight, where everything in nature seems reduced to a single shade of aqua. At the same time, our peak sensitivity shifts from yellow-green to blue-green, a change first reported by nineteenth-century Czech physiologist Jan Purkinje, the guy who also first suggested that fingerprints might be useful in solving crimes. We still call that low-light alteration the Purkinje shift. This human sensitivity to green explains why the US interstate highway system, initially constructed in the 1950s, uses mostly green signs and why increasing numbers of municipalities now purchase green instead of red fire trucks.

As light fades further, even green-blue hues vanish. Our eyes then work solely by their twenty million rod-shaped retinal cells. It's as if our eyes have switched to a different kind of film. Our scotopic vision is dead-drunk slow to get going. Rod cells are lazy;

they need repeated stimulation to operate at all. At night, when you turn off your bedroom lights, you first see nothing but total darkness. Within a few seconds some details slowly emerge. After five minutes the general features of the room are apparent. In twenty minutes, if you're still awake, you see everything you're ever going to see. But if someone clicks on the bright lights for a moment, then switches them off, you're back to where you started, and you must begin the process of "dark adaptation" from scratch.

Our scotopic vision is also color-blind. That red sweatshirt and those blue socks you threw over the chair—they're gray now. The whole room is monochrome. (Incidentally, the only animals known to be totally blind to all colors are owls—suggested by the fact that their feathers are not vividly hued, a telltale sign that they have no need to use colorful plumage in order to attract the opposite sex—although their dim-light monochrome acuity is almost infinitely sharper than ours.) Unlike other colors, deep reds don't just go gray in dim light; they vanish altogether. Rod cells simply cannot perceive wavelengths longer than 630 nanometers, which, unfortunately, includes the most common color in the universe, the red glow of excited hydrogen that is the calling card of giant interstellar gas clouds like the Orion nebula. One fascinating demonstration involves Christmas lights on a rheostat, or dimmer switch. As the brightness is turned way down, blue, yellow, orange, and green bulbs reach a dim point where they go gray. But deep red lights never turn gray. When they get sufficiently faint, they simply vanish.

If you're a human—a likely state of affairs if you're reading this book—your night vision is very blurry. Normal vision is said to be 20/20 in bright light, although many young people could read the 20/10 line on the Snellen chart if only they were asked to try. (That's line number 11, the bottommost.) But *in dim light our best visual acuity is 20/200.* That's legally blind. If you're going out with someone on a first date and are strolling along a dark street or in a park at night, you could play the sympathy angle by telling him or her that you are legally blind. You wouldn't be lying.

In dim light you have a blind spot in the very middle of your field of vision. It's located dead center, it's present in both eyes, and it's large—twice the size of the full moon as it appears in the sky. This happens because only cone cells are located centrally, which is why we see best in bright light by looking directly at the object of interest. But at night we are best able to perceive faint details when looking slightly off to the side. Astronomers have known this for centuries. Most eyes can resolve the many individual stars in the Beehive star cluster in the constellation Cancer, but only by using *averted* vision. When stared at directly, the cluster is a blurry blob.

Understanding scotopic, or faint-light, vision provides keys to understanding our blindness to many kinds of invisible light as well. Every night of our lives, we find ourselves blind to colors that we effortlessly see in the daylight. So when you're thinking about the invisible part of the electromagnetic spectrum—radio waves, microwaves, infrared light, ultraviolet light, X-rays, and gamma rays—you can put them in the same category as color frequencies that are undetectable to our night vision. They're there, no less real than they are during the daytime; we just can't see them. The reason is threefold.

First, outside the visible wavelengths and infrared radiation (which we *do* perceive, or at least our skin does, as heat), we enter the realm of energies, such as microwaves, that the sun either does not emit at all or gives off very weakly. Why should our vision be able to locate objects only when microwaves bounce off them? The sun emits almost no microwaves, and thus these objects are not around us in nature. Why should we have a visual architecture that can detect things that aren't there and that don't affect us?

Second, retinal cells consist of molecules that are affected by incoming energies, which trigger the cells to emit electrical pulses. The energy of photons of various kinds determines how the retina interacts with matter. It's no accident that retinal photochemistry only works when dealing with a very limited range of energies. If the spacing between waves of light is just a bit longer than that of the deepest red we can see, its energy is too weak to influence the retina's protein molecules. This is why, on a strictly physical and chemical level, we cannot see infrared radiation.

At the other end of the spectrum, the energy of wavelengths shorter than those of violet is so powerful that it can damage those sensitive molecules. Fortunately the lens of the eye absorbs this light before it can do any retinal damage. (The lens pays a price for shielding the retina, though. Over time, the lens itself becomes damaged, often resulting in cataracts. This is why it's a good idea to block the blue-violet-ultraviolet part of the spectrum by wearing sunglasses in bright natural light.)

Third, and perhaps least important, some forms of invisible light don't bounce off the objects around us. Some of these wavelengths bend or diffract around objects instead of reflecting off them. Others, such as X-rays, penetrate objects rather than bouncing off them. Detecting these rays as a way of "seeing" our environment would be impossible and pointless.

Now that we know what they are and why we can't see them, let's begin our tour of the invisible rays that flood our lives.

CHAPTER 3

The Green Planet and the Red Heat

Ours is an age of discovery. A time of mind-blowing scientific revelations, ranging from magnetars (bizarre supermagnetic suns) to extremophiles (life forms that thrive in seemingly lethal temperatures). We're used to being wowed by science.

So it's hard for us to imagine the shock that swept the planet in March of 1781, when an unknown astronomy hobbyist abruptly pulled the rug out from under the greatest minds of the time, achieving instant global fame.

Who was this person whose name was then on everyone's lips? Odds are you've scarcely heard of him. Indeed, when we ponder the greatest minds in human history—those whose discoveries were most profound or ahead of their time—we'd have to begin with an "unknown": Aristarchus of Samos, the first person to say that the earth orbits the sun. We'd include Isaac Newton, who explained motion, and Albert Einstein, who revealed that space and time warp and shrink, meaning that the universe doesn't have a fixed size. And among these giants we must place a man named William Herschel, born on November 15, 1738, in Hannover, Germany. He may not have been as brilliant as the aforementioned pioneers, but thanks to his tenacity—his ability to follow a scientific scent as relentlessly as a bloodhound—he made not one but *two* of the most astonishing scientific discoveries of his time. On top of that, he was just plain lucky.

Herschel's early life gave little clue to his later genius. He was one of ten children, and his father was not an aristocrat but rather a musician in the military, an oboist. Following in his father's footsteps, young William played oboe in the band of the Hanoverian Guards, where he displayed enough talent at an early age to suggest that composing might be his lifelong destiny.

He first visited England at the age of eighteen and was so impressed that he resolved to immigrate there, which he did the following year. Moving to a new place to seek one's fortune as a musician seems like a Haight-Ashbury pipe dream in our modern times, but back then it was an almost mainstream thing to do. England during the mid- and late eighteenth century was a land of musical opportunity like no other, and Mozart, Haydn, and Handel, along with thousands of unknown musicians, gravitated to the country. As a result, the competition was fierce. Yet Herschel soon scraped together a living by copying music in that pre-Xerox era. Slowly he climbed the professional ladder, paying the rent by teaching and composing. Finally, in 1766, he was appointed organist of a fashionable chapel in Bath, the well-known spa town. Herschel was not only an accomplished oboist and organist: he was also skilled on the violin and harpsichord. He composed twenty-four symphonies and many concertos—an impressive output, even if the works are regarded today as remedies for insomnia.

Happily, his intellectual curiosity extended far beyond music. One day he read a book that changed his life: Robert Smith's *A*

Compleat System of Opticks, from which he learned techniques for do-it-yourself telescope construction. With his boundless energy, Herschel didn't merely mimic the astronomers of his day by observing and sketching the moon and planets. He was primarily intrigued by the dim nebulae that peppered the heavens. Almost universally believed to be luminous fluids, they beckoned with a mystery unrivaled by the mountains and craters of the moon. What were these hazy blobs, really? They were so faint that to see them well would require a telescope larger than any then in existence, since the brightness and clarity of a telescope image was directly proportional to the diameter of its main lens, or mirror. The big problem, then as now, was that bigger telescopes were disproportionately expensive, and many were of poor quality.

If Herschel wanted a mirror big enough to see the nebulae up close, he was going to have to make it himself. Back then, specialized mirror glass didn't yet exist, and Herschel began casting molten metal in his basement and grinding his own mirrors from blank disks he created by mixing copper, tin, and antimony. Although his first mirrors cracked on cooling and became nothing more than expensive paperweights, he ultimately succeeded, and indeed his mirrors grew ever larger, some nearly two feet in diameter, a huge light-gathering improvement from the mere six- to eight-inch mirrors that were most common at the time. More than that, their shapes—paraboloids fashioned to a precise curve—were of such high quality that all the distant starlight, planet light, and galaxy light arrived at a perfect common focus. Herschel's homemade telescopes outperformed even those at the famous Royal Observatory, in Greenwich. He also made his own eyepieces.

Herschel was joined in this work by his brother Alexander and his sister Caroline, who remained his faithful assistant for the rest of his long life and eventually became a respected scientist in her own right.

Slowly the intellectual society of England began to hear whispers of this unusual family with their enormous, unparalleled telescopes. Using them, Herschel embarked on perhaps the most energy-intensive project of his life—a telescopic survey of the entire sky. When that was completed, he built an even bigger telescope and completed an even more thorough and detailed survey of the heavens. Then, on March 13, 1781, during his third and most comprehensive sky survey, he saw something that would change his life and astound the world.

Herschel observed a green "star" that wasn't just a point of light but was shaped like a disk. Initially he surmised that it must be a comet. But it couldn't be a comet, for it never developed a tail or the highly elliptical orbit comets possess. By observing the object's slow nightly motion, he soon realized it was a new planet that had an eighty-four-year orbit around the sun. He named it Georgium Sidus after the English king George III, in a successful attempt to attract attention and curry favor. However, others soon insisted that it be named in accordance with tradition and possess the name of a god in Roman mythology. Thus the first planet anyone *ever* discovered came to be called Uranus.

The world was amazed. No one had imagined that the universe contained any planets other than the five familiar bright "wandering stars," which had been known since prehistory and were mentioned in the Bible, the Vedas (the Hindu holy books, written in Sanskrit), and ancient Egyptian papyri. No prophet, no holy book, no great thinker, no prestigious council, no school of philosophy—no one had entertained the notion that there could be other worlds that were too faint to be readily seen. Moreover, the telescope had been around for 170 years by that point. Countless astronomers had carefully inspected the heavens. The cosmos contained one sun, one moon, and five bright planets. That was the nature of reality, and no one had the slightest reason to doubt it until Herschel blew that "reality" to smithereens. The shock of the discovery was akin to a modern scientist revealing that it is our largest toenail, rather than the brain, that controls our thoughts. Or that the moon is hollow and inhabited by a race of monkeys.

As it turns out, Uranus is dimly visible to the unaided eye—I've seen it on several occasions without the slightest optical aid. That green world is positively brilliant through any telescope. Why hadn't anyone found it earlier? This question caused deep consternation among intellectuals. It became the headline of the century around the world.

Overnight William Herschel went from an amateur telescope maker and mediocre classical composer to the world's most celebrated scientist. The Royal Society awarded him the Copley Medal, the era's equivalent of the Nobel Prize. King George III, badly needing some prestige after having just lost possession of the American colonies and tickled pink that Herschel had first tried to name the new planet George, awarded him an annual pension of two hundred pounds sterling.

Herschel could then spend all his time on astronomy, and he did so tirelessly for the next forty years. He dedicated his new career to building ever-larger telescopes and trying to solve the thorny nebula problem—deciphering the true nature of those interstellar "clouds." He almost immediately found that his best instruments rendered most of those glowing blotches as separate innumerable stars, which made him wrongly conclude that all nebulae were of that nature—star clusters. But since some nebulae stubbornly remained blurry no matter how large the telescope or great the magnification, he knew they must be very distant as well as huge. He therefore concluded that the whole cosmos was composed of such gargantuan clusters—cities of suns—which would later be called galaxies.

In 1788 the Herschels moved to Slough, where William spent the rest of his life. On every clear night (in England such nights were few and far between; he hired a watchman to wake him if the weather cleared), Herschel observed the heavens, dictating from the eyepiece while Caroline took notes. To supplement his income, Herschel made telescopes—the finest of his day—for others.

When all was said and done, Herschel created three catalogs listing 2,500 nebulae and star clusters. He identified 848 double stars and had seventy scientific papers published. He correctly calculated our solar system's motion and direction through space. He discovered two moons of Saturn and coined the word *asteroid*. He was the first to say that our Milky Way galaxy has the shape of a pancake. Not bad for an oboist.

But Herschel appears here not for his first-ever discovery of a new planet but for something he found in 1800, near the end of his illustrious career. Actually, his unveiling of the green planet was so astounding that this later revelation is scarcely mentioned in biographical sketches. *Encyclopaedia Britannica*'s 1,700-word article about him devotes exactly ten of those words to his having discovered the first-ever invisible light.

Though he was the first to discover them scientifically, Herschel was not the first to *intuit* the existence of unseen rays. But he, like everyone else, was unaware of previous hypotheses because the work of his predecessor had evaporated decades earlier, leaving no trace. Émilie du Châtelet, who died in 1749, was a French author, physicist, and mathematician. Her greatest accomplishment remains her translation and explication of Isaac Newton's main work, his *Philosophiae Naturalis Principia Mathematica,* a translation still in widespread use today in the French-speaking world. Du Châtelet had been raised in Paris in the 1710s, in a thirty-room town house overlooking the Tuileries Gardens. As a child she loved eavesdropping when educated guests—especially astronomers—came to visit, and she developed a talent for science and mathematics that was rare among the women of her time, as few women had the opportunity to study these subjects. (The writer Voltaire later said that she was "a great man whose only fault was being a woman," which underscores the virtual impossibility of gaining respect or recognition as a female scientist in that era.)

Émilie was in her late twenties when she met Voltaire, then age thirty-nine, and they fell in love. Soon they were living together in a large house in eastern France, where together they created a center for research—she in science, he in philosophy. Distinguished visitors regularly came from other parts of Europe.

One summer night, Émilie had an insight about the nature of light that would reverberate more than a century later in the wake of discoveries involving photography and infrared radiation. She wrote her ideas in accurate scientific form and, with a bit of influence from Voltaire, became the only woman in eighteenth-century France to have a scientific treatise published. That 1737 paper, *Dissertation on the Nature and Propagation of Fire,* ultimately published in 1744, predicted that there was an unseen form of light that she hypothesized must be the source of a flame's heat.

After she and Voltaire broke up, she fell in love with a French poet, and when their relationship ended—badly—she discovered that she was pregnant with his child. In that era, a pregnancy in one's forties was extremely risky, and Émilie had a premonition that she would die in childbirth. Frantically she rushed to complete her magnum opus—the translation of and commentary on Newton's *Principia.* Then, tragically, her premonition came true, and she did die soon after childbirth, at age forty-two. Just as sad, a commotion erupted when word of her unwed status spread a scandal that quickly cast her life, work, and accomplishments into oblivion. She remained unknown well into the twentieth century and didn't really emerge until a 2006 revival of interest resurrected her accomplishments. In any case, her prediction about an invisible component of light was of course utterly unknown to Herschel a lifetime after her published paper had appeared and just as promptly vanished from sight.

Fully nineteen years after his discovery of Uranus brought him global renown, Herschel was still engaged in nonstop experimentation and observation, always recorded by his sister Caroline (continuing the pattern of women making unsung contributions to Herschel's discoveries). Neither he nor anyone else realized it, but an avalanche of publicity and acclaim was about to come crashing down on him all over again.

In 1800, Herschel was fully aware that when visible light strikes any surface, some of the light's energy is absorbed, and the surface is warmed. He knew that dark objects apparently absorb more energy than light objects, since they get hot faster, whereas white paper, for example, reflects most of the light that hits it and scarcely warms at all.

Herschel was then using two of his telescopes to observe the sun. He employed dark glass filters to screen out most of the light so it wouldn't blind him. Yet he could always still feel some of the sun's warmth coming through the filters. He noticed that some filters seemed to allow more of the light to pass through, while others transmitted more of the heat. He wrote that, when observing the sun, he "felt a sensation of heat" even when a particular filter transmitted "but little light; while others gave... much light with scarce any sensation of heat."

Always curious about the underlying nature of things, Herschel decided to see for himself the degree to which various colors of glass transmitted heat. Were some colors associated with more heat than others? It was the simplest kind of question, but it was one that nobody had previously explored.

With Caroline looking on, pen in hand, Herschel set up an apparatus that let sunlight pass through a narrow opening. The thin beam of light then struck a piece of cut glass. This prism spread a rainbow containing all the colors of the visible spectrum across his table. He then positioned three thermometers on the table. He placed two in the shadows, far outside the spectrum. Their purpose was to act as "controls," taking the temperature of the unlit sections of the table. Then he placed the third thermometer within the bands of colored light on the table, studying each color in the spectrum one by one.

Herschel took repeated readings in the violet, green, and red regions of the spectrum. In each he observed a temperature rise, which he dictated and Caroline recorded. His findings: after he left the thermometer for eight minutes in each of the three colored-light zones, the average temperature increase was two degrees Fahrenheit in violet, 3.2 degrees Fahrenheit in green, and 6.9 degrees Fahrenheit in red. The red was much hotter than the other colors!

Obviously either sunlight's red beams had a greater heating effect than its green and violet beams or perhaps there was somehow *more* red light reaching the table than green or violet light, although this latter explanation didn't seem likely because the red *looked* no brighter than the other hues.

Then something historic happened. It was the kind of fluke we frequently come across when we study the great eureka moments of scientific discovery. Herschel took a break from his experiment and left the room. As the sun moved slowly across the sky, the spectrum crept across the table until it was no longer hitting any of Herschel's carefully placed thermometers. When he returned, he glanced at the thermometer he'd left sitting in red light, by then lying in the shadow just outside the red end of the spectrum, and was surprised to see that the temperature was much higher than it was when the instrument had been bathed directly in the sunlight's red band. What was going on? He repeated the readings. The conclusion slowly dawned on him. Invisible "heat rays" coming from the sun were being refracted by the prism to a position just beyond the rainbow's red boundary.

Herschel kept taking measurements at various positions. If he placed the thermometer more than a few inches past the red end of the visible spectrum, it registered no temperature increase at all and matched his two control instruments. He also looked beyond the violet end of the spectrum but found no temperature change there.

William Herschel, who made not one but two astonishing discoveries, including that of the first-ever form of invisible light. *(Wikimedia Commons)*

Herschel published three papers in one of the Royal Society's journals, reporting his results. He also quickly discovered through further experiments that this same invisible heat-light also emanated from terrestrial sources such as gas lamps and candles. At the end of his second paper, Herschel suggested that light and heat are part of the same phenomenon, tracing two apparently disparate experiences to the same source. "We are not allowed, by the rules of philosophizing, to admit two different causes to explain certain effects, if they may be accounted for by one," he wrote.

Here Herschel was using a scientific principle already known in his day called Occam's razor. It was posited by an English Franciscan friar, William of Ockham (1287–1347), who argued that of all the possible explanations for a given result, the one with the fewest assumptions is most likely to be correct. In other words, when it comes to scientific hypotheses, the simpler the better. If your car fails to start in the morning, it's possible that the problem was caused by a meteorite falling on it overnight and damaging the electronic ignition circuit board. But it's simpler to surmise that the battery is dead or you're out of gas. If these simple assumptions turn out to be wrong, then you might gradually move on to investigate increasingly intricate and unlikely scenarios.

So Herschel could have imagined that his prism bent, or refracted, two entirely different solar phenomena and placed them side by side on his table. But it was simpler—and, as it turned out, correct—to assume that light contains one component we see with our eyes and another that we feel as heat on our skin, heat that can be measured with a thermometer. He also suggested that light's various colors might produce dissimilar effects on chemicals and their reactions, foreshadowing the birth of photography half a century later.

Herschel called the new, invisible form of light he had stumbled upon *calorific rays* and announced that these rays were reflected, refracted, absorbed, and transmitted just as visible light is. (We will remember this similarity later, because other forms of unseen energy do *not* follow this pattern and do *not* behave as visible light does.) Despite deriving from the Latin word for "heat" (*calor*) and thus being a logical label, the term *calorific rays* was later abandoned, and Herschel's invisible rays were renamed *infrared radiation* to indicate their location in the spectrum. This, of course, was logical, too, because the prefix *infra* means "below" the red.

Herschel himself suggested that if we could detect these rays in the distant universe, they might open the door to new discoveries that would remain unavailable if we limit ourselves to the visible alone. This proved true as well. Today more than half of all new telescopes are designed to detect infrared rays, including the giant James Webb Space Telescope, scheduled for launch late in 2018. This is one of the reasons modern observatories are located atop tall mountains, where our atmosphere will absorb and block the least amount of cosmic infrared radiation. When the European Space Agency launched its orbiting infrared telescope, whose 3.5-meter mirror makes it much larger than the Hubble Space Telescope (though smaller than the Webb), they named it the Herschel Space Observatory. They also took pains to explain that it was named to honor both William and his sister Caroline. It detects radiation from objects in space that are too cold to emit visible radiation.

And so the infrared energy discovered by William and Caroline lit the way, so to speak, for an ongoing flood of discoveries. It also very much paved the way for the discovery of an entirely different form of invisible light—one that was downright spooky.

CHAPTER 4

Hot Rays

Radiation that lies just beyond the red end of the spectrum may be invisible, but it has a major effect on us. First things first: we must stop ourselves from making the mistake everyone does—regarding infrared radiation and heat as the same thing. We cannot be blamed for this, since everyone calls a deep-red bathroom floodlight a heat lamp. And on a chilly April afternoon, we might claim to "feel the sun's heat" on our skin. Neither is true. In both these cases, what's being emitted is infrared radiation.

Infrared radiation is not heat. Rather, infrared radiation *creates* heat. They are two entirely different animals. You are now probably the only one on your street who knows this.

How does infrared radiation create heat? Think of infrared radiation as invisible waves of light, a millimeter from crest to crest, generated the same way other forms of light are created, by the motion of electrons. Heat, on the other hand, is simply the motion of atoms and molecules. Little solid particles moving or jiggling. When a light ray's waves have just the right spacing so that they pass a particular spot at the correct frequency, it will give an entire atom a little push. Beam enough of those waves at anything, and all its atoms will be set to vibrate. Voilà: heat.

Visible rays of sunlight can jostle atoms, too, as Herschel discovered. But if the waves are too short and too frequent, they'll scarcely budge atoms, which is why green and especially blue and violet light don't heat things very much. Red does better. And when it comes to maximum efficiency, those unseen infrared waves are optimal for making whole atoms jiggle.

Say you're sitting in a parked car. Visible sunlight comes through the glass. Some of it manages to slightly jiggle some of the atoms in the dashboard, upholstery, and every other sunlit part of the auto's interior. Then each jiggling atom creates and releases some of its newfound kinetic energy as a bit of infrared light, so now the car's interior is increasingly filled with infrared light. And here's the surprise: while glass is transparent to visible rays (duh!) it is opaque to infrared rays. That's because the regular resonance, or vibration rate, of glass atoms is naturally in sync with infrared rays, so a kind of chaotic barrier gets created, preventing the infrared rays from leaving the car. Glass lets visible rays in but keeps infrared rays from exiting. *Glass creates an infrared trap*. Result: the car gets hotter and hotter. This is why you never leave a living thing in a closed car on a sunny day.

And it's why greenhouses, sometimes called hothouses, are made of glass. We want to let light in but block heat from escaping.

When you feel the "warmth of the sun" you are actually only feeling the increased vibrations of your skin's atoms. The accelerated atomic motion was *caused* by the sun's infrared radiation, but you can't actually feel it. The sun does a good job of warming your skin, the ground, anything it hits, because slightly more than half the sun's total emissions are pure infrared radiation. There's lots of it hitting our planet. When it strikes the ground it heats it so efficiently that the surface of the earth in turn heats the air just above it. Hot air rises, so up go bubbles of warm air. When this starts happening, typically around 9:00 or 10:00 a.m. every sunny day between late April and late August, the air rises far enough—cooling steadily all the while at the rate of five degrees Fahrenheit per thousand feet—to eventually reach a height where it cannot hold its vapor, since cool air can't contain as much moisture as warm air. Bingo: a cloud suddenly forms. The water has changed from its gaseous state, which is transparent, to its liquid state, meaning that the vapor abruptly condensed into untold billions of tiny droplets. There are around

A cloud is a visible manifestation of the power of infrared rays. The sun's IR radiation heats the ground, which makes a large air bubble rise until it cools too much to hold its vapor in a gaseous state. *(Bob Berman)*

five grams of water droplets in each cubic meter of a typical cloud. No wonder you can "feel" that you're in a cloud whenever you walk through fog. It feels viscerally damp.

The cloud's height reveals the altitude at which the rising air has cooled to its dew point, where the vapor-to-liquid transition occurs. In humid summer air, that's typically four thousand feet. In the dry air of Montana it's more like nine thousand feet. Infrared radiation is what kicks the daily weather into motion.

Another way infrared radiation differs from heat is that the latter spreads rather slowly, while infrared radiation travels at the speed of light, essentially instantaneously. Remember that heat is the motion of atoms. So let's say we place a frying pan on the stove and turn on the burner. The flame makes the pan's metal atoms jiggle faster and faster as the bottom heats up. But the atoms in the pan's handle are still cool. You can hold it safely for quite a while, even as the butter starts melting and the eggs start frying. The handle will eventually heat up, though, because heat always travels in a single direction, from hot areas to cool areas, which means that the kinetic energy of fast-moving atoms influences other, more stationary atoms in their vicinity and makes them start to jiggle faster, too. The jiggling domino effect gradually moves through the metal until at last the handle gets too hot to touch. It takes a while.

But now consider infrared radiation, which, again, moves at the speed of light. Say you're gathering with some friends around a bonfire. Even at a distance you can feel the fire's heat on your face, which actually means you're feeling the flames' infrared rays making your skin's atoms move faster. But here's the point. If some big person steps in front of you, blocking the light, you instantly feel the change on your face. The guy in front of you has cast an infrared-blocking shadow; you are now inside it. *You can feel the infrared's absence immediately.*

Where there's heat, there's infrared radiation, which makes it a boon to police, the military, and weather reporters. Infrared light makes atoms jiggle, and the energy of jiggling atoms causes new infrared photons to appear. In every situation except on the hottest summer days, bodies are warmer than their surroundings, so your jiggling atoms emit infrared rays and betray your presence to infrared sensors. Such devices can also detect various crops (think marijuana), because each type of plant has an

Cloud layers hover at specific altitudes because the temperature at those elevations causes atmospheric moisture to change from a vapor to liquid droplets. The unseen energy behind all this drama is infrared. *(Michael Mah)*

exact characteristic temperature. On a more benign note, infrared-sensing satellites can determine the heights of clouds, because the top of a cloud is cooler than the bottom.

Such orbiting instruments can also pinpoint ocean movements, especially dramatic ones such as the frigid Humboldt Current, which runs northward up the western coast of South America.

Because infrared waves bend around obstacles more easily than visible light does, they're used in some cordless phones and audio headsets. They've been controlling garage doors for decades, sending invisible rays to the door opener's IR detector, which signals the motor to start moving. Medical infrared imaging is a very useful diagnostic tool, too, because tumors, for example, are usually warmer than their surrounding tissues. Infrared cameras are effective fire-detecting tools because they can spot heat alterations in buildings and can be used to test electronic fire-prevention systems.

But the best thing about infrared waves, what sets them apart from some of the other invisible rays we'll explore, is that they're harmless. Infrared rays are not carcinogenic, and their ability to heat skin does not pose a mortal risk. After all, it's okay to have your body atoms jiggling a bit faster than usual. Indeed, when you get the flu and run a fever, all your atoms are moving around three miles per hour faster than normal. You could tell the doctor that your atoms are so fast right now that they're making you feel terrible, and he might give you some paracetamol and say, "Here: this will slow them down." The point is that a little extra atom speed isn't going to hurt you very much. (There are exceptions, or course. People and pets have died of sunstroke.)

But compared to more closely spaced, faster-vibrating light rays such as gamma rays and X-rays, which, as we'll see, can be seriously harmful, long waves such as infrared rays are far more benign. Not only are they safe, they're also essential to your coziness, since infrared radiation creates heat. And we have infrared radiation to thank for other daily comforts and conveniences as well.

And okay—it may not be technically correct, but go ahead and keep referring to that bathroom infrared floodlight as a heat lamp. At least now you know the real story.

CHAPTER 5

Ultraviolet Brings the Blues

Our next invisible light "flavor" is ultraviolet. This is an attention grabber, because it's the variety of light that causes the most deaths every year. Its discovery was troublesome, too, and carries us back more than two centuries, to a mere year after William Herschel announced his calorific rays to the cheers of scientists around the globe.

It's hard to imagine a story more different from Herschel's than that of Johann Ritter. For introducing the world to infrared radiation, Herschel was justly lionized, almost worshipped, and lived happily into his mid-eighties. By contrast, Ritter's 1801 discovery of ultraviolet light was ignored—and he died penniless at the age of thirty-three. It hardly seems fair.

Ritter was born in Samitz, in what is now Poland, in 1776—five years before Herschel discovered Uranus. The son of a Protestant pastor, he attended grammar school and, at age fourteen, was sent by his father to a nearby city to learn to be a pharmacist. As he was learning and practicing the trade, he developed a keen love of science and performed countless experiments. His life changed, however, when he was nineteen: his father died and left him a modest inheritance, enabling him to enroll in the

University of Jena in April of 1796. There he met Alexander von Humboldt, one of the most famous scientists of his day, and embarked on studies of electricity and its effects on the body.

By the time he reached the age of twenty-one, Ritter's research and published papers on electrochemistry and electrophysiology had attracted enough notice to make him an odds-on favorite for the ranks of the science immortals. Indeed, after he created the world's first dry-cell battery, at the age of twenty-four, then married the young, pretty girlfriend with whom he'd been living for years, his trajectory seemed upward bound.

But it was not to be. Ritter had a tendency to get into disputes, and he quarreled with university officials over the issue of whether he'd be appointed as a lecturer. In time, Europe's larger scientific community also grew suspicious of Ritter's conclusions and even many of his stated facts because he routinely couched them in philosophical or occultist rants.

The main problem was his championing of the tenets of the *Naturphilosophie,* then in vogue in some German intellectual centers. Ritter based his work on the idea that the universe is a "oneness" in which all scientific disciplines are interconnected and that the cosmos possesses a "world-soul," meaning a kind of innate, built-in, God-like intelligence. He also believed that *polarity*—meaning the law of opposites—ruled nature. Just as magnets have a north and south pole, Ritter was convinced that absolutely every aspect of science boiled down to, and could be explained in terms of, pairs—the elements of which, in his view, often stood in opposition to each other. He found evidence for his belief everywhere he looked. One of his experiments refined the work of earlier scientists who'd used electricity to divide water into two elements, hydrogen and oxygen. He also noted that air had been found to be essentially composed of two gases, oxygen and nitrogen. Always pairs. He became convinced that the earth had opposing electrical poles in addition to opposing magnetic poles.

It was his obsession with dualities that spurred Ritter to his greatest discovery. He had of course learned of Herschel's bombshell 1800 discovery of invisible calorific rays, or heat rays, just beyond the red end of the spectrum. Ritter quickly hypothesized that *cooling* rays might dwell on the opposite side of the spectrum, the violet end.

He started out by doing exactly what Herschel had done, but he soon found that temperatures did *not* drop on the violet end, so he tried something else. Since he couldn't find any physical effect produced by the rays, he looked for a chemical reaction. It had already been proved that paper soaked with silver chloride would blacken when exposed to sunlight; this discovery was one of the earliest stepping-stones toward the field of photography. Ritter wondered whether all sunlight's colors would create this reaction with equal speed. He exposed silver chloride–soaked paper to various parts of the prismatic spectrum, cast onto the paper by sunshine striking cut glass. Red light had only a negligible effect in darkening, or reducing, the compound to silver, while green light did it much faster and violet did it fastest of all. Ritter then placed the chemically soaked paper in the dark spot beyond the violet end of the spectrum, and voilà—the paper darkened even more rapidly than it had in violet light. Obviously some invisible rays that lay beyond the violet end of the spectrum had a dramatic, repeatable chemical effect.

Ritter had done it—he had discovered an entirely new form of invisible light. But alas, he predictably interpreted this effect as proof of a polarity between "deoxidizing rays" near the violet end of the spectrum and "oxidizing rays" near the red end. Here again was his obsession with dualities. When the world heard about it, people soon abandoned Ritter's term "oxidizing rays," and this new form of invisible light came to be called *chemical rays,* a label that stuck throughout most of the 1800s. It took a full lifetime for Herschel's *calorific* and Ritter's *chemical* rays to instead be labeled infrared and ultraviolet.

You'd think such a momentous finding would have elevated Ritter to the status of Herschel. It didn't. First Ritter continued his habit of embedding his findings in the language of polarities and soul, the *Naturphilosophie* tenets of "oneness between nature and people." And soon he got worse, routinely peppering his papers with references to such occult practices as dowsing (using a divining rod to find water underground). He imagined he'd found the general principles governing the interdependencies of inorganic nature and human phenomena and named this new branch of study siderism. He even published a periodical with that title. With virtually no subscribers, its first issue was its last.

Perhaps worst of all, Ritter took forever to publish anything. After making a discovery or conducting an original experiment—and some were truly exciting and profound—he'd make a brief, cryptic, and vague announcement in some science journal, then wait years before taking the trouble to explain what he'd found. But he would only write about it in his books, where descriptions of his results were tangled up with extraneous stuff about supernatural phenomena. He once confessed that after two months of experiments and discoveries, it took him two years to adequately write it all up.

This was far too slow, especially at a time when science was advancing at an accelerating clip. Ritter was doing important work, but his advances in the study of galvanism and electricity were delayed in reaching Europe's science community, and even when they did arrive, they were met with skepticism.

All in all, Ritter wrote thirteen volumes about his scientific findings, books that told of several groundbreaking discoveries, including the effects of electricity on animal bodies and, of course, that first-ever dry-cell battery. He also published twenty journal articles. Yet despite all this work he remained generally unknown and failed to land a single teaching appointment. Increasingly in debt, unable to care for his family, and in failing health, he died of pulmonary problems just after his thirty-third birthday.

His is the saddest story of all our invisible light discoverers. For decades after his death, few people had heard of him, especially outside Germany. He remains generally unknown even today. But while Ritter lived and died in obscurity, the ultraviolet light he discovered gained increasing attention. For as the world soon learned, UV is the invisible radiation most intimately involved with human life—and death. Its rays will powerfully influence your health.

CHAPTER 6

Danger Beyond the Violet

When it comes to ultraviolet light's effect on humans, we are in a bit of a bind: we can't live without it, but too many unfortunates each year find that they can't live with it, either.

Ultraviolet light's power comes from the sawtooth fineness of its waves. Herschel's infrared pulses are separated by as much as a millimeter—approximately the diameter of a needle. They pulse a trillion times a second. By comparison, ultraviolet waves are just one two-millionths of an inch apart. Their separations are submicroscopic. At least a thousand trillion (one quadrillion) of them flash past you every second. Light with longer wavelengths—such as infrared radiation, microwaves, radio waves, and visible colors—can jostle entire atoms, and they may even be able to slightly heat living tissue, but they are too weak to interfere with atomic structure. The lightning speed of ultraviolet light, on the other hand, gives it the power to strip electrons from atoms. As a result, atoms are broken apart—*ionized.*

It's this ionizing power that makes UV light so dangerous. If an atom that's important to our health—such as any of the atoms that make up our DNA—is ionized, the damage can be lethal, inducing gene and cell mutations that are the precursors to cancer. More than eight thousand people in the United States are killed annually by melanoma (skin cancer), nearly all of which is caused by exactly this process. On the other hand, the vitamin D the body produces when struck by ultraviolet light may be the most potent cancer-preventing substance known. In 2016, a major medical journal announced that UV exposure helps prevent pancreatic cancer.

Just above our atmosphere, 10 percent of the sun's energy is in the ultraviolet band of the spectrum. But air is so effective at blocking it that fully 77 percent of the UV rays trying to penetrate it fails to reach the earth's surface. Thus sunshine measured on the ground has a maximum ultraviolet component of just 3 percent. The rest of it consists of 44 percent visible light and 53 percent infrared radiation. Moreover, the intensity of UV radiation at the earth's surface varies throughout the day (it is greatest when the sun is directly overhead) and with the seasons (it is greatest in summer).

All ultraviolet light is not created equal. Ninety-five percent of the UV light striking us is UVA, the weakest variety and the one with the longest waves—from 3,200 to 4,000 angstroms—some of which is actually visible to the eye as violet. If you let a prism break sunlight into its rainbow spectrum, the most violet-looking color at the extreme fringe, where the color dims a bit before reaching its blank endpoint—that's UVA. UVA has been scientifically associated with malignant melanoma, but in general it's the safest variety.

Slightly shorter wavelengths ramp up the danger big-time. If you could actually see the type of UV light whose waves are spaced 3,000 angstroms apart, it would look virtually identical to the UV light whose waves are 3,200 Å apart. Yet that 3,000 Å variety, with waves just slightly closer together, creates sunburns eighty times faster!

UVC, the most powerful and lethal variety, is defined as UV light whose waves are between 2,800 Å and 400 Å apart— incredibly close. UVC is not only cancer-causing, it also has the power of speedy sterilization. Fortunately our atmosphere is so effective at blocking UVC that only a single UVC photon reaches the surface of our planet every thirty million years. So we don't have to worry about UVC endangering our health. (Those who are gung-ho about future human colonization of other planets, however, should definitely read the UVC product manual, because every conceivable astronaut-visitation destination, including Mars and the moon, gets continuously bathed in this ultrapowerful UVC.)

Public enemy number one is UVB (whose waves measure between 3,200 and 2,800 Å in length), the villain responsible for sunburn and various skin cancers. This is the Goldilocks UV—its wavelengths are long enough to sometimes get through our atmosphere but short enough to have ionizing power. On the other hand, UVB is also best at inducing production of vitamin D in the skin, at rates of up to 1,000 international units *per minute*. That's amazingly fast production, which is good for us, because it means we don't have to stay exposed to UV rays for too long to get what we need.

Around 1 percent of the light hitting a sunbather is UV light—more if the sun is high overhead. That may sound like very little, but in fact it translates into a million trillion photons of UV rays per second. All are capable of altering DNA. Epidemiologists have discovered a disturbing link: every 1 percent increase in lifetime UV exposure produces a 1 percent boost in skin cancer incidence, although the truly deadly skin cancer, melanoma, seems to have its origin mainly in severe sunburns rather than ordinary sun exposure. Obviously the bottom line is: avoid sunburns.

That's harder than it sounds. UV is a subtle beast, and it's important to arm yourself with knowledge of its quirks so you can protect yourself. Lounging under a beach umbrella isn't good enough, because one-third of UV radiation scatters in the atmosphere, so that it comes at you sideways. Because of this atmospheric scattering, half the UV reaching you comes not straight from the sun but rather from the sky's UV brightness—utterly unseen by the eye. Relaxing in the shade is inadequate protection against sunburn if you're exposed to a big swath of sky, which is why summer beachgoers who think they've been careful often return home with bright red shoulders and faces. Your surroundings strongly influence your likelihood of tanning or burning, regardless of whether you're in sun or shade. Twelve percent of UV light reflects away from dry sand, but just 5 percent reflects away if the sand is wet, so the choice of where you spread your blanket can make a big difference in whether you get burned. Plopping yourself by the high-tide mark, with the wet sand and surf nearby, will give you less of a burn. Vegetation absorbs nearly all UV radiation and essentially reflects nothing, certainly less than 10 percent in all cases. So a lawn picnic will expose you to far less UV than an equal amount of time spent at the beach, even if it means choosing ants over sand flies. In addition, over the course of your lifetime, you may have noticed that you burn more easily on windy days. That's because water generally absorbs UV light when it's calm but reflects it when it's rough or rippling. Thus a lakeside or riverside picnic is far safer on a calm day than a breezy one.

Thanks to dense air along the horizon, a very low sun sends us reduced visible light as well as reduced infrared radiation, and its ultraviolet emissions are essentially at zero. Unless you are at a high altitude, you will almost never get a tan or burn during the final two hours of daylight. *(Bob Berman)*

We may associate UV exposure with beach days, but snow reflects six times more UV rays than even dry sand does. Because a whopping 80–90 percent of the ultraviolet light striking snow is reflected, you'd think skiers and snowboarders would all get burns. Well, sometimes they do and sometimes they don't. It depends on when they decide to hit the slopes.

The time of day and the sun's elevation determine how much ultraviolet light arrives to cause trouble. As the sun loses height, its light has to pass through an increasing amount of air. That's because the lowest layers of the atmosphere are denser, or thicker, than those above them, and when we sight low we gaze exclusively through these low layers. To use real numbers, compared with an overhead view, we look through twice as much air when we view a star that's one-third of the way up in the sky (thirty degrees high) and fourteen times more air when we look at the setting sun. And because air filters out UV rays, the more there is of it, the fewer UV rays can get through. From late April through mid-August, then, the sun is so high between 11:00 a.m. and 3:00 p.m. that you can burn in an hour or less.

In spring and summer, your safe, low-UV hours are between 5:00 a.m. and 9:00 a.m. and between 5:00 p.m. and 8:00 p.m. During those seven daylight hours UVA intensity is more than halved, and UVB is chopped by at least 80 percent compared with midday exposure. The sun may look and feel strong, but you won't burn. Skin will temporarily redden because infrared light reaches you at nearly full strength, but the UV light is too weak to harm your skin unless you are very fair. So if your family is a tribe of blue-eyed blondes or redheads who all burn at the slightest provocation, the safest outdoor summer activity—no hat or sunscreen required—is a late afternoon picnic in a green environment: grass adores UV and sucks it in, reflecting a mere 3 percent toward your skin.

Back to our snowboarders. Because the sun stays low in the sky between November and February, your UV exposure during those months is minimal, even if you're outside all day. But the sun's midday elevation starts climbing rapidly in February, rising by the width of two suns (or what appears to be the width of two suns) every week. Beginning in mid-March, and especially during late March, the midday snow reflects prodigious amounts of UV light, which explains why late-season snowboarders do indeed get badly burned. Snow-reflected UV rays in March produce a burn three times faster than they do in December.

A hazy day screens away half the incoming UV rays. When the air feels very humid, it takes twice as much time to tan or burn. A cotton T-shirt or undershirt blocks 90 percent of the harmful UV rays coming at you, but the best UV-blocking fabrics are tight weaves such as denim. Less important is the color: nonetheless, a bright Day-Glo yellow will block more UV radiation than a muted color such as gray will.

Single-pane window glass blocks half the UV rays hitting it. Standard double-pane windows block even more. If it takes you one hour to tan outdoors, it will take fifteen hours behind a typical window. So feel free to indulge in all the nude indoor sunbathing you want (as long as you don't have creepy neighbors). The more UV-reducing factors in play, the safer you are: sitting in a greenhouse in winter when the sun is low, you'd need at least 160 hours of steady noontime exposure to burn.

Travel alters UV exposure, too. It's greatest at low latitudes because of the sun's high elevation every day of the year, and it's greatest in low-humidity environments, such as most of Australia. You may not have known that simply going to a place at a high elevation will crank up your UV exposure big-time. Every thousand-foot climb raises UV intensity by 4 percent. That's why if you're in Leadville, Colorado, you'll get hit with 40 percent more UV than if you're in Washington, DC, though both sit at the same latitude and thus receive equal solar intensity.

It will come as no surprise that clouds play a big role in filtering

UV rays. "A thickly overcast day means no UV at all," NASA atmospheric physicist Jay Herman wrote me in an e-mail. "It's why Australians with their sunny climate have such a high rate of skin cancer compared with most Americans." The clouds' thickness matters greatly. Heavy, dark clouds may block all UV, but a high, thin cirrus layer lets almost 100 percent of the UV that reaches it pass through.

Our greatest protection from UV rays lies between six and thirty-five miles over our heads, in a layer of pale blue gas composed of molecules made of three oxygen atoms apiece. Peaking at an altitude of fifteen miles, the ozone layer is the earth's primary shield against UVB. But it's an amazingly delicate barrier: if all the air's ozone were to settle at the earth's surface, it would only be as thick as two stacked pennies.

Although ozone-destroying chemicals such as CFCs, used in aerosol sprays and refrigerants, and halon, used in fire extinguishers, have been banned, those long-lived molecules will continue to cause damage for a few more decades. NASA physicists monitor UV radiation with special satellites and predict that UV-blocking ozone will return to normal levels by 2050. Meanwhile, people with fair skin should, unlike mad dogs and Englishmen, stay out of the midday sun. For the rest of us, each day's variable solar conditions require a judgment call. And probably sunscreen, whose SPF (sun protection factor) numbers reveal how effective the product is. A lotion with an SPF of 10, for example, tells us that if current conditions would give you a burn in one hour, the sunscreen will forestall the burn for ten hours.

Most sunscreens use either titanium dioxide or zinc oxide, chemicals that reflect, scatter, or absorb ultraviolet light and

The solar wind, a continuous flow of charged particles released in the sun's upper atmosphere, excites air atoms as close as one hundred miles up from the earth's surface. The result is the beautiful glow of the aurora, seen here from central Alaska. But every night, no matter where you are, though the pattern is much dimmer and more diffuse than it is during the daytime, the sky gives off a visible radiance—the effect caused by ultraviolet rays from the sun exciting atoms in our atmosphere. The sky always glows! This is why in rural places, far from artificial lights, nocturnal hikers can still see well enough to follow a trail, except where overhead foliage obstructs the sky. *(Anjali Bermain)*

dissipate it as heat. In practice, creams labeled SPF 30 provide 96 percent of the protection of an SPF 90 product, which suggests that you'll do fine with a 30 as long as you keep reapplying it as needed during the day.

If you're one of the many millions who spend long hours working indoors, you may think that because you spend very little time exposed to the sun you can wash your hands of the whole UV issue. But it's not that simple. UV radiation doesn't just threaten us; it also sustains us and has the power to heal us. And our way of life is making it increasingly difficult to get the UV we need.

CHAPTER 7

Energy Rhythms

We used to take the cycles of day and night for granted. We obeyed them; we had no choice. But times have changed.

Sure, everyone knows that our planet spins once around its axis in twenty-four hours—even if the true figure, if we use the distant stars as a reference point, is twenty-three hours, fifty-six minutes, and 4.1 seconds. And everyone knows that all diurnal animals undergo waking and sleeping cycles in sync with a natural rhythm of light and darkness. But we also assume that humans are somehow above it all. Just as in the 1940s, when most sophisticated urban women in Western countries regarded the act of breast-feeding infants as somehow primitive and unnecessary and believed that babies raised on powdered or bottled formula would do just as well, so, too, did we imagine that we could dispense with being a slave to daylight. As Shakespeare's Juliet said, "All the world will be in love with night and pay no worship to the garish sun."

The shift away from regular sun exposure began in the nineteenth century, as the United States and parts of Europe began a steady transition from mostly outdoor, agrarian societies to mostly indoor, industrialized societies. Accelerating this trend was the rise of the night shift in many areas of commerce during the early twentieth century, when Thomas Edison's durable lightbulb allowed factories to run their assembly lines 24-7. At the same time, increasing numbers of people started to take night courses at colleges that offered them. We had become a society in which a hefty minority of us stayed awake during the night and slept during the day. Aside from some yawns at awkward times, there seemed to be little consequence. As usual, we were wrong.

In the midst of this massive shift to nocturnal, indoor activities, people who worked during the day shifts also started to avoid direct sunlight—on purpose. It wasn't the visible rays they were shunning but the invisible ultraviolet rays. One of the deadliest forms of cancer, melanoma, was on the rise by the late 1950s, spurred by an uptick in the popularity of outdoor sports, such as fishing, golf, and tennis, and the start of a mass migration to sunny cities in the American Southwest. Even though it claimed only around eight thousand lives annually, melanoma developed into a major media narrative that quickly turned into full-blown heliophobia.

Exacerbating the fear was the fact that far more common and much less aggressive skin cancers were indeed becoming endemic—but as far as media reporting was concerned, cancer was cancer. It can be deeply disconcerting to hear your physician tell you that you have a tumor. His or her next sentence, explaining that it is merely a benign variety that cannot metastasize and is therefore unlikely to cause serious harm, scarcely detracts from the fact that you've heard the dreaded word.

Studies showed that skin cancers of all types were far more prevalent in places where the sun was strongest. Moreover, people who spent the most time outdoors—e.g., on boats or playing golf—experienced the highest incidence of skin cancer. This was more than enough information for the general public as well as the medical community. The new anticancer suggestion was: avoid the sun.

Thanks to modern architecture, it was easy for people in the 1960s to do just that. Increasingly, buildings were constructed with windows that could not be opened—and glass effectively blocks the sun's invisible ultraviolet emissions. When out of the office, people rode in cars, where the availability of air-conditioning starting in the early 1960s encouraged closed-window driving. Then in the 1980s a new product appeared on the market: sunblock, which essentially stops the body's vitamin D production cold. In their early years, such lotions—with numbers like SPF 30 and SPF 45 and names like Coppertone, a product that was introduced in the early 1950s—had been primarily marketed as tan-promoting products. But with the advent of the late twentieth century, sunblock was sold as a skin-cancer preventive, and people were advised to cover themselves with it throughout the summer months. Even the medical establishment urged hiding from the sun as a way to avoid skin cancer.

At the same time, a series of major developments altered children's exposure to ultraviolet rays as well as to visible solar rays. Until the 1970s, kids were encouraged to play outdoors after school. They'd hit the parks, ball fields, and playgrounds and materialize at home just before dinnertime. All that changed with, first, the widespread fear of crime, especially sexual predation. Next came the video-game craze in the 1980s, which meant that kids were opting to shut themselves in their rooms rather than climb trees. In the 2000s, texting and Internet surfing came along to increase indoor time for kids even more. By the early years of the twenty-first century, the metamorphosis was complete. Humans in Western countries had transformed themselves into a race of mole people. Avoidance of the sun was almost total. At the same time, levels of vitamin D in blood plunged to virtually zero. Nowadays, many vitamin D researchers and expert groups say that a blood level of at least 30 ng/mL (nanograms per milliliter) is optimal; some advise even higher goals—40 or 50 ng/mL.

In the midst of the great indoor migration, we apparently forgot something humans have known for millennia. The ancient Greeks, who worshipped the sun god Helios, seem to have been the first to write about the importance of sunlight in human health. They of course didn't know that it's the UV component of sunlight that produces its most salutary effects, but nonetheless the benefits of sunbathing appeared in the writings of Herodotus (fifth century BCE), Cicero (first century BCE), the architect Vitruvius (first century BCE), Pliny the Elder (23–79 CE), the famed Roman physician Galen (130–200 CE), the Greek surgeon Antyllus (second century CE), and others.

After the fall of the Roman Empire, the practice of sunbathing apparently fell into disfavor. But it emerged again during the early Middle Ages, as documented by the Persian scholar and physician Avicenna (980–1037 CE). Sunbathing for medical and cosmetic purposes has continued ever since. Cultures around the globe maintain a belief in the healing power of sunlight. Or at least we did until the 1980s, when skin cancer worries abruptly changed the picture.

It may be true that excessive sun exposure increases risk of skin cancer. But paradoxically, it also seems that sunlight can lower our cancer risk. The late Dr. Robert Heaney of Creighton University, in Omaha, who treated thousands of patients with vitamin D and was a member of the nonprofit Vitamin D Council, pointed out that vitamin D was found to be beneficial in thirty-two randomized trials. In one big study of women whose average age was sixty-two, a large daily vitamin D supplement produced a whopping 60 percent reduction in all kinds of cancers after just four years, compared to a control group.

Heaney was not alone in believing that vitamin D prevents tiny, predetectable cancers from growing and spreading. "That's the kind of cancer I'd want to have—one that never grows," he told me.

In any discussion of the natural rays, visible and invisible, in which our bodies evolved, obtaining sufficient UV-provoked vitamin D_3 makes much sense medically. (There are five chemically distinct types of vitamin D. The one the body creates when struck by ultraviolet light is also the kind that, when taken as a supplement, is most associated with reduced mortality, especially in the elderly. This is vitamin D_3.) After all, vitamin D is automatically created in our bodies when our skin is struck by the sun's ultraviolet rays. Why would our bodies rapidly create it if it weren't important to have in our blood? Ten minutes in strong sunlight stimulates your body to create as much vitamin D as you'd get from drinking two hundred glasses of milk. The natural conclusion we can draw from this is that the human body needs to have a high and steady level of this vitamin in circulation. Yet if we hide from the sun, we'll suffer a deficiency.

As for the fear of skin cancer, an unexpected revelation comes from Dr. Stephanie Seneff, a senior scientist at MIT who has been conducting research on the relationship between nutrition and health for decades. "Both cholesterol and sulfur afford protection in the skin from radiation damage to the cell's DNA, the kind of damage that can lead to skin cancer," Dr. Seneff told me. "Cholesterol and sulfur become oxidized upon exposure to the high-frequency rays in sunlight, thus acting as antioxidants to 'take the heat,' so to speak. Oxidation of cholesterol is also the first step in the process by which cholesterol transforms itself into vitamin D_3."

According to Dr. Seneff, our bodies contain the necessary mechanisms to extract or produce beneficial nutrients from the sun while also shielding us from harm. "Circumventing this natural process, either by using sunblock or staying out of the sun entirely, makes us lose all the health benefits and gives a variety of disease processes free rein," says Dr. Seneff.

Dr. John Cannell, of the Vitamin D Council, summed it up for me in a few words: "Everyone should get as much sunlight as they can, without burning."

In addition to its cancer-preventing value, sunlight plays a critical role in preventing or curing rickets, a terrible disease that mostly affects children. Rickets is caused by the absence of vitamin D, which in turn causes dietary calcium to be inadequately absorbed, resulting in skeletal deformities and neuromuscular symptoms such as hyperexcitability.

Sunlight's ultraviolet rays are also associated with mitigating depression. These days we use the term SAD to denote seasonal affective disorder, a severe depression linked to the reduction in sunlight we experience during the winter. It afflicts 15 percent of the population, and it begins and ends at around the same time every year. Says the Mayo Clinic on its website: "If you're like most people with SAD, your symptoms start in the fall and continue into the winter months, sapping your energy and making you feel moody." The primary treatment for SAD is light therapy, or phototherapy—deliberate daily or near-daily exposure to natural or artificial light. The most commonly used phototherapy product is a "light box," but before you step into one, talk the matter over with your physician. If you suffer from both SAD and bipolar disorder, for example, then a too-quick increase in light exposure could induce manic symptoms.

Now that you know about the health benefits of both the visible and invisible components of sunlight, let's explore its opposite: darkness. For our ancestors, nightfall ushered in a period of total darkness, illuminated only by moon, stars, and fire. Now many of us spend the night awash in the glow of computer screens, alarm clocks, and streetlights. But just as your body's need for adequate sunlight is important, some studies suggest that our bodies also need to be *shielded* from the electromagnetic spectrum, including rays emitted by indoor light.

Here we come to an astonishing medical revelation: the single surest cause of breast cancer is—a lack of darkness!

According to the Breast Cancer Fund, "Effects of night-shift work on breast cancer risk are greatest for women who work rotating hours that include the overnight (as opposed to evening) shift and for those who work twelve-hour shifts that frequently switch between day and night work. . . . Risk of developing breast cancer increased for women who worked night shifts for more than four and a half to five years, especially those who regularly engaged in night work for at least four years prior to their first pregnancy, [i.e.,] before their mammary systems had fully differentiated."

These results are concerning, the fund explains, because around 15 percent of the US workforce currently works at least some of the time on non-day shifts. The report continues: "The most thoroughly studied mechanism to explain the effects of night-shift work is called the light-at-night (LAN) hypothesis. Increasing exposure to light, especially bright indoor lights, at times outside of normal daylight hours, decreases secretion of melatonin," which can increase the risk of breast cancer. "In support of the light-at-night hypothesis, blind women who are completely unable to perceive the presence of environmental light, and therefore have no daily decreases in melatonin levels, have a statistically significant lower risk of diagnosis of breast cancer than do blind women who do perceive light and have regular decreases in melatonin secretion over the normal twenty-four-hour cycle. The former effect (no daily melatonin decreases) and its opposite in night-shift workers (no daily melatonin increases) both support the conclusion that the greater the secretion of melatonin, the lower the risk of breast cancer."

This all suggests that human health not only requires *visible* sunlight but also very much demands the *invisible* portion of sunlight—ultraviolet light. And, perhaps almost as important, the body must be in tune with sunlight's natural cycle, which means that we must also regularly experience its absence.

The takeaway here is that artificial indoor light contradicts nature's alternating dark-light rhythm. Artificial light may be fine for extending our hours of daily productivity, but it shouldn't be left permanently on during our sleep cycle. In other words, we need darkness. What remains unclear is the quantity: how much dim light is sufficient to create a medical problem? Is it the amount given off by a night-light? A clock's LED digits? A

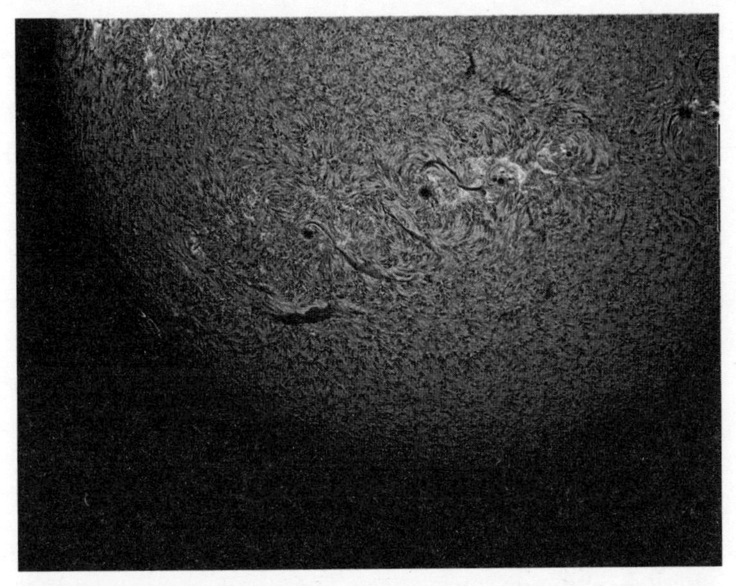

Ultraviolet light from the sun stimulates the human body to create vitamin D, one of the most potent anticancer agents known. Research suggests that people should not routinely use sunscreen to block ultraviolet rays but actually expose themselves to as much sunlight as they can without burning. *(Matt Francis, Prescott Observatory)*

streetlight shining behind closed curtains? Passing car headlights? How much is too much? At the time of this writing there's no clear-cut answer.

Besides its health benefits, UV light helps us see things that are invisible to us in regular light. The reason we can't see UV light is because its waves are absorbed by the lens of the eye and don't reach the retina. Good thing, too, because if they did reach the retina, they would damage the cone-shaped receptor cells responsible for color vision. But some animals, including reptiles, birds, and many insects, including butterflies and bees, *can* see UV rays. This isn't accidental. Many seeds, fruits, and flowers "pop" in UV light and are easier to find against a busy background. Famously, scorpions glow brilliantly under UV illumination. Some birds even display plumage designs that are invisible in ordinary light. Moreover, the semen and urine of many animals, including humans, will glow in UV light. Health inspectors use a source of UV light along with detection equipment to spot unclean and improperly washed hotel rooms. That's because bodily fluids fluoresce: their complex chemicals absorb UV light and emit it as visible light.

Suburbanites commonly use UV light fixtures to attract flying pests that "see" its waves, luring them to "zappers" or traps. These devices emit the kind of UV light that has the longest waves—UVA, just barely unseen by the human eye—because flies are most attracted to light at 3,650 angstroms—right in the middle of UVA's range.

In addition to killing bugs, ultraviolet light helps us understand the cosmos. The ultraviolet universe has a very different appearance from familiar galaxies and stars, which mostly glow in visible and infrared light. That's because most stars are cooler than our sun, and even our sun only emits 10 percent of its light as UV rays. Indeed, 95 percent of all stars shine overwhelmingly in the infrared and visible parts of the spectrum, and the majority emit more infrared and less visible light than our own beloved sun does.

Ultraviolet radiation—detection of which is performed by orbiting spacecraft above our UV-blocking atmosphere—is always the telltale fingerprint of unusually intense heat. Superhot stars, which appear blue to the eye, are either anomalously massive or are being viewed in the early or very late stages of their evolution, when their electromagnetic emissions are unusually intense. Such unusually high heat does more than crank up the emissions across their entire spectrum—it also shifts their emission curve to the shorter, more energetic end of the spectrum, making blue light and especially ultraviolet light more plentiful.

Instruments that detect ultraviolet light in the heavens do not show most stars at all. But they do show the violence of merging galaxies and the extreme temperatures of the giant blue stars that even in visible light often serve as the lighthouses of the cosmos. In short, UV telescopes offer a portrait not of average, middle-mass, ordinary stars like our own sun but rather of the exceptional places where violence is routine. Through UV "eyes," it's always the apocalypse.

The awakening of science to this new way of seeing the cosmos began with Johann Ritter in 1801. By 1815, scientists found that Ritter's "chemical rays" darkened not just silver chloride but also many other kinds of metallic salts. Between 1826 and 1837, Nicéphore Niépce, credited with taking the first successful photograph, in 1827, and Louis Daguerre, the most famous photographic innovator of his day, found that silver iodide was especially light sensitive, and they used this discovery as the basis for their early work, which even then had begun to gain international notice. By 1842, others found that when sunlight hit a gelatin emulsion containing silver iodide, soon to be called a daguerreotype plate, it induced a photochemical reaction. Practical photography was born.

During the remainder of the nineteenth century, physicists kept making important theoretical and empirical discoveries that clarified the nature and properties of UV rays, although they were still called chemical rays until the 1870s. The development of very bright artificial lighting such as carbon arcs—dazzling spotlight-type fixtures that emit light by causing high-voltage electrical charges to leap across a short gap—provided the world with light sources that gave off copious UV radiation. The most important breakthrough came in 1859, when Gustav Kirchhoff and Robert Bunsen invented the spectroscope, which shows the composition of light by splitting it into its constituent wavelengths. After that, scientists could identify any light-emitting substance simply by noting the patterns of colors in its light. As they watched a burning building in the distance, for example, their spectroscope told them that the pipes contained lead. Suddenly the fields of physics, astronomy, and chemistry changed forever. The spectroscope enabled researchers to determine the composition of anything—even a star—merely by observing its emitted light.

Years later, scientists learned that the sun's light—visible and invisible—is merely the by-product of a process alchemists had vainly tried to reproduce for centuries—the transmutation of one element into another. That nature accomplishes this before our very eyes, and that it is what creates the solar heat and light that supports all life, was suspected by no one. The revelation came as a complete surprise.

Let's temporarily leave our timeline of invisible-light discoveries to explore this astounding central talent of the sun.

CHAPTER 8

The Exploding Sun

In Woodstock, New York, aging ex-hippies of my acquaintance still say things like, "It's all energy!" They're right. While old-time physicists used to speak of frictional energy, chemical energy, mechanical energy, electrical energy, kinetic energy, potential energy, and lots of other varieties, science has come to see that, almost mystically, "it's all one!"

Meaning, first of all, that the universe was born with all the energy it will ever have. And what it has never diminishes. This seems counterintuitive, because we do use up the fuel in our cars—and watch some of it get wasted as heat that goes out through the exhaust pipe or as heat that the rubber tires leave on the road. Energy seems to diminish. But in reality, it merely changes form.

Let's look at the forms of energy that used to seem so distinct and various. Say you slam on the brakes in your car, an action that requires mechanical energy and that uses up chemical energy in the form of gasoline. Your car then slows down because frictional energy on the asphalt and the brake pads causes the tires to turn more slowly. All sorts of different energies seem to be required simply because the idiot in front of you screeched to a halt the instant the traffic light turned yellow.

But look more closely. The mechanical energy involved moving parts, as did the frictional energy. Friction converted the mechanical energy to heat—and what is heat? It's merely the motion of atoms. That's all there is to it—the simple movement of atoms and molecules. So really what happened was that a macroscopic (visible) display of energy—tires slowing—was changed to a submicroscopic form, as umpteen atoms were sped up. In total no energy was lost. It just changed form. And it involved nothing more than motion of various kinds. So it turns out that all energy is motion.

Other examples? Well, if we start with thermal energy, or heat energy, all superhot objects routinely convert their heat into visible electromagnetic radiation (light). Or you can use a thermocouple to convert thermal energy into electrical energy. Or employ a steam turbine to convert thermal energy into mechanical energy.

Or you could go the opposite way and start with mechanical energy, such as the kind found in an engine. You could convert it into a different form of mechanical energy via gears or levers. Or you could convert it into nuclear energy via a synchrotron or particle accelerator. Or into thermal energy by hitting your brakes. Or into electrical energy by using a generator. Or into chemical energy by striking a match.

It's clear that any and all energy can change forms. But energy can never be destroyed or used up. Scientists nowadays say that all energy falls into one of two categories—either *kinetic* energy (the energy of motion, including those scampering atoms that comprise heat) or *potential* energy (meaning money-in-the-bank future energy, such as the kind stored in a car parked on a hill, which can release its brakes anytime and glide down on the strength of its favorable position in a gravitational field).

The sun illustrates energy conversion beautifully. Technically it's converting nuclear energy to electromagnetic energy. Put another way, it's obeying Einstein by changing its mass to energy, as expressed in the famous equation $E = mc^2$. Before anyone figured this out (it was Arthur Eddington, in 1920), the sun's prodigious light and heat were an utter mystery. Science had already calculated that a massive ball of coal with the sun's weight—the mass of 333,000 Earths—would completely burn itself out in two thousand years. But the sun is obviously older than this, so it just couldn't be burning in the usual sense.

It wasn't. Instead the sun's high internal heat means that its hydrogen atoms move furiously enough to smash together. When four of them combine, they create a single atom of helium. That's the whole story.

It so happens that a helium atom weighs just a smidgen less than four hydrogen atoms, so there is a loss of mass in this fusion process. The mass is released as its energy equivalent. Using Einstein's equation, the conversion of a single pencil eraser's worth of mass to energy could light up all the electric bulbs in the United States for thirteen days. In the sun, the conversion involves four million tons of hydrogen per second. That has a bit more bulk than a pencil eraser, so the resulting energy output is staggering.

This is not some theoretical figure. If we had a giant scale and could weigh the sun, our nearest star, we'd find that it actually weighs four million tons less every second. We might get worried and say, "Whoa! Slow down!" But given that the sun has a total mass of two nonillion—that's the number 2 followed by twenty-seven zeros—tons, its ongoing loss of mass is not noticeable. It'll be billions of years before any serious consequences ensue.

The sun's tiny fusion reactor, a small ball at its exact center, which has just one two-hundredth of the volume of the sun in its entirety, emits mostly gamma rays and X-rays as a by-product of that hydrogen-to-helium conversion. But the energy keeps getting absorbed and reemitted as it tries to squirm its way out. After as many as a million years, the original ultra-high-energy photons are now safely shifted down to a mix of visible light and infrared light with a small percentage of ultraviolet light thrown in. This final mixture—the sun's rays—leaves the sun's visible gaseous surface, the photosphere, at a speed of 186,282 miles per second, then takes a mere eight minutes to reach us and deliver that energy.

So our lives depend on the solar mass-to-energy conversion. The result is very nearly an equal mixture of invisible light and the rainbow colors. The sun's blue light then gets scattered around by our planet's atmosphere, giving us our blue sky. So when we look at the sun it seems yellowish—the result of its missing the blue that went into painting the daytime sky.

The sun's fusion reactor, that unseen ball at its center, is more than the source of all life. The conversion of one form of energy to another is what the universe does, inside and outside our bodies, here on earth and in every cosmic village.

CHAPTER 9

No Soap

It's a good thing we can't see radio waves. They'd overwhelm us. They're absolutely everywhere, bouncing off layers of the atmosphere above us and reflecting off hills and buildings. They pass through our bodies as if we were ghosts.

Being longer than visible light waves, they're also wimpier. In fact radio waves are the weakest kind of light. But their impact on our world has been anything but weak. Once they were discovered, radio waves were rapidly integrated into our technologies—a process that continues today, well into the twenty-first century. This most harmless kind of invisible light is also the kind that has most changed our daily lives and continues to do so. But don't imagine that radio waves are some kind of sound. Sound is merely a mechanical compression, usually of air, that sets our ear drums vibrating. By contrast, radio waves are a form of light whose waves are each much longer than those of the visible colors.

The brilliant Heinrich Hertz is justly credited for first discovering radio waves and demonstrating how they work. But he couldn't have done it without the genius groundwork of two men who came before him.

The first player in this particular tale is Michael Faraday, born in England in 1791 in what was then Surrey and is now the London borough of Southwark. He received minimal formal education, and in his youth he showed no sign that he would become one of the most influential scientists in history. A century later, Albert Einstein kept only three pictures on the wall of his study to serve as inspiration. They were portraits of Isaac Newton, Michael Faraday, and James Clerk Maxwell—who, oddly enough, is the other main character in our radio story.

During his seven-year apprenticeship to a bookbinder, Faraday came across a single volume that changed his life: Isaac Watts's *The Improvement of the Mind*. Now largely forgotten, it was an early self-help book that contained a list of sixteen general rules for the improvement of knowledge. It also contained an overview of books and reading, a guide for study and meditation, tips on improving one's memory, and other suggestions for the betterment of the self. Faraday happened to read it just as he was developing a fascination with science, particularly the then-mysterious field of electricity, and he embraced the book's principles and suggestions.

The next critical turn in Faraday's life came as he completed his bookbinding apprenticeship, at the age of twenty. Like today's youth, who sometimes feel a bit adrift after obtaining their undergraduate degrees, Faraday still wasn't sure what he wanted to do in life, so he checked out the science lectures being given in his neighborhood. He was particularly drawn to talks by the famous chemist Humphry Davy, of the Royal Society. Faraday was so enraptured by everything he heard from Davy that he eventually sent his idol a three-hundred-page book of the notes he made from Davy's lectures. Anyone would have been flattered by such a gift, and Davy immediately responded with a grateful and kind reply. The following year, when Davy damaged his eyesight in a lab accident involving the highly unstable compound nitrogen trichloride, he sent word to Faraday, offering him a job as his assistant, which the latter immediately accepted.

Thus began a series of fortuitous events that was to change the course of science. Soon one of the assistants at London's Royal Institution—like the Royal Society, an organization devoted to scientific research—was fired, and the task of finding a replacement was handed to Davy. By then he was so impressed with Faraday's genius and initiative that in March of 1813 he appointed Faraday the organization's chemical assistant. Later that spring, the understandably NCl_3-shy Davy started letting Faraday prepare the nitrogen trichloride samples. It was a prudent step, but it still proved to be no insurance against lab disasters. Sure enough, both men managed to get injured in yet another explosion of this dangerous compound.

Even so, their relationship thrived. In late 1813, Davy planned to begin a one-and-a-half-year professional tour of Europe, and he asked Faraday to accompany him as his assistant. Unfortunately, Davy's valet resigned just before the tour began, so Faraday was asked to assume that menial role as well. One hopes that Davy was respectful to his colleague during this time, but it can't have been easy for Faraday. English society was extremely class-oriented in the early nineteenth century, and Davy's wife, Jane Apreece, never treated Faraday as a peer. She sent him to eat with the servants, and even when there was plenty of room in the coach, Jane insisted that he ride outside, no matter how hard it was raining. Faraday became so disheartened that he came very close to abandoning the whole thing, returning to

England, and switching from science back to bookbinding. But he stuck it out, which is how he got introduced to many of Europe's top-notch scientists, their keen minds, and their groundbreaking ideas.

When he returned to England, Faraday began researching the magnetism that shifted compass needles in the vicinity of wires carrying electricity. He found that simply moving a magnet through a loop of wire produces an electric current. Alternatively, he could create a current by moving a loop of coiled copper around a stationary magnet. His experiments showed that a changing magnetic field produces an electrical field—a relationship mathematically articulated three decades later by James Clerk Maxwell, who called it Faraday's law. Perhaps the most amazing of Faraday's revelations was the concept of an *electromagnetic field,* its unseen curving lines extending into empty space. This spooky and original idea—a force field penetrating the unoccupied area around electrical wires—was initially rejected by other scientists. Still, Faraday's notion of fields emanating from objects that have electrical and magnetic charges, as well as the implication that electricity and magnetism are inextricably linked as a single entity, proved to be spot-on. He showed that magnetism can affect light (although never measurably altering its path) and suggested that light, too, is a manifestation of magnetism and electricity.

Faraday even built the first crude motors using the interplay of wires and magnets. His was the first power generator, and his discoveries form the basis of all the electric motors that dominate modern technology and our everyday lives. Beyond the gasoline-powered engine, every automobile also uses more than a dozen electric motors that owe their origins to Faraday: the electric starter, the power windows and doors, the push-button seat adjustments, the windshield wipers and washers—all of them are electric. Their increasing use starting in the 1990s resulted from the development of ever-more-powerful magnets based on rare earth elements, such as neodymium, europium, and yttrium, which provide a high level of torque in small spaces. An iPhone, for example, uses eight rare earth elements.

Faraday lived into his mid-seventies, honored, world renowned, but always remarkably humble. He twice refused when he was asked to be president of the Royal Society. When the queen offered to knight him, he turned down the chance to be Sir Michael: he believed it was against the spirit of the Bible to pursue riches and worldly rewards. He preferred, he said, "to remain plain Mr. Faraday to the end." And he declined on ethical grounds when the British government requested his advice on the production of chemical weapons for use in the Crimean War.

Faraday died in 1867. Two years earlier, a Scottish physicist expressed Faraday's observations in mathematical terms, proving that Faraday was right. Well off, educated, and brilliant, James Clerk Maxwell displayed an intense curiosity about nature even as a young child. According to his biographer Basil Mahon, when Maxwell was as young as three years old "everything that moved, shone, or made a noise drew the question: 'what's the go o' that?'" Maxwell's father, in an 1834 letter to his sister-in-law Jane Cay, describes the boy's innate sense of inquisitiveness: "He has great work with doors, locks, keys, etc., and 'show me how it doos' is never out of his mouth. He also investigates the hidden course of streams and bell-wires, the way the water gets from the pond through the wall."

His curiosity never ebbed. Maxwell spent his life solving

such seemingly intractable problems as the nature of Saturn's rings. (He mathematically showed why they could be neither a single solid structure nor a liquid nor a gas but must instead be composed of innumerable separate tiny rocks—*moonlets.*) He essentially established the basis for color photography with his scheme for creating negatives that would reproduce individual primary colors. But it was his lengthy 1861 paper, *On Physical Lines of Force,* that set the stage for his most important work: four differential equations, forever after called Maxwell's equations, that mathematically explained electromagnetism, showing that oscillating electrical and magnetic fields travel through space as waves and move at the speed of light. Maxwell's equations also showed that radio waves should exist. This logical unification of magnetism, electrical fields, and light is what inspired Heinrich Hertz's search for, and discovery of, the existence of radio waves exactly twenty-one years later. Who can say what else Maxwell might have discovered had he not succumbed to abdominal cancer at the age of forty-eight?

Heinrich Hertz, born in 1857 in Hamburg into a prosperous family, was fascinated by Faraday's experiments and Maxwell's mathematical conclusions about them and methodically set out to see if he could demonstrate the existence of these putative electromagnetic waves. Like Maxwell, Hertz would die young—the result of a failed operation for an infection—at age thirty-six. His life was short, but his work affected nearly every life that came after his own.

Hertz's groundbreaking 1886 experiment involved the first ever "spark gap transmitter," which used wires to carry current but left a gap in the circuit so that one could observe the current as it leaped across empty space. These energies were neither infrared nor ultraviolet nor visible. They were something else—yet they had the power to create a visible electrical "snap" through empty air.

High-voltage sparks like these are accompanied by unseen radio-wave emissions. *(Kevin Smith, www.lessmiths.com)*

Hertz calculated that each of these mysterious waves was around thirteen feet from crest to crest, roughly the length of a small car. He also measured the electrical-field intensity, polarity, and reflection of the waves from solid surfaces, particularly metal ones. To him these waves appeared to be a form of electromagnetic radiation that perfectly obeyed the Maxwell equations.

Remember the two basic properties of electromagnetic waves mentioned in chapter 1? *Wavelength* is the distance from one peak of a wave's electrical field to the next peak. The *frequency* is the number of these waves that pass you every second. Maxwell had shown that those two basic properties are intimately related and inversely proportional. Meaning that if a wave is huge—say, the distance light travels in one second, or 186,000 miles from crest to crest—then just a single one will pass by per second. But if a wave's length is one thousand times shorter—in this case 186 miles from crest to crest—then its frequency will be one thousand times faster. In that case, one thousand such waves would pass you per second. Conclusion: wavelength (in miles) times frequency always equals 186,282—the distance in miles that light travels in a second. Bottom line: the shorter the wavelength, the faster the waves must pulse. Moreover, the shorter the wave and the faster its frequency, the more energy it has.

Read the preceding paragraph one more time so that it is absolutely clear, and you will have grasped the most important properties of all light, visible and invisible.

The world soon honored Heinrich Hertz by naming the unit designating a wave's frequency the hertz. A wave that pulses 700 times a second is said to vibrate at 700 hertz. If it pulses 700,000 times a second, we use the prefix *kilo* (meaning "one thousand") and say it has a frequency of 700 kilohertz, or 700 kHz.

If you have an old AM radio, the kind with numbers on the dial, take a glance at it. The station frequencies are listed in kilohertz, or thousands of pulses per second. If your favorite AM station is at 710 on the dial, it means that 710,000 of its waves zoom through you and your radio each second.

In the FM band, waves are much shorter and thus pulse much faster. FM frequencies are expressed in megahertz, or millions of pulses per second. If your favorite FM station is 90.1 on the dial, then around 90 million waves pass by per second.

An AM station of 1,000 (kilohertz) would be identical to an FM frequency of 1 megahertz, because a thousand thousands are a million. Turns out, a 1,000 kHz wave has a length of 299.8 meters, or 984 feet. That's around a fifth of a mile, or four city blocks. This 984-foot distance from one wave crest to the next happens to be one-millionth the distance light travels in one second, so one million of these waves must zoom past you per second. In short, roughly speaking, electromagnetism with a thousand-foot wavelength has a frequency of a million hertz. That's the story for a radio frequency of 1 mHz, or 1,000 kHz. I hope you're taking notes on all this.

If we instead consider a wave one hundred times smaller, meaning ten feet from crest to crest, its frequency must be one hundred times faster, or 100 megahertz. A hundred million of these ten-foot waves pass by per second.

This frequency—100 megahertz—happens to be commonplace in the FM broadcasting universe, so it's worth visualizing. Picture a series of waves, each ten feet across. Now picture 100 million of them flashing past you, and through you, every second. Good luck.

Hertz's groundbreaking work, "Researches on the Propagation of Electric Action with Finite Velocity Through Space," announced his discovery of this new form of electromagnetic radiation—radio waves. But he saw no practical benefit to them whatsoever; he seemed satisfied to have proved their existence, and his inquiry ended there. His students soon got wind of all the attention the discovery was bringing to their professor. They'd ask him what it meant, and he'd say, "It's of no use whatsoever. . . . this is just an experiment that proves Maestro Maxwell was right—we just have these mysterious electromagnetic waves that we cannot see with the naked eye. But they are there."

On another occasion, a noted physicist asked him what technological value these waves had. His reply was short:

"Nothing, I guess."

But others knew better. It may have required two decades between the time Maxwell laid out the mathematical reality of electromagnetic waves based on Faraday's findings and Hertz's proof that they really exist. But it took a mere eyeblink for enterprising inventors to put them to use.

CHAPTER 10

Turning On and Tuning In

The idea of waves traveling through the air, and the possibility that they could be focused or bounced off surfaces, made inventors salivate. Some envisioned sending telegraph signals consisting of dots and dashes—short and long pulses—without the need for wires and poles. Such wireless communication between land and ships at sea, using what were at first called Hertzian waves, or aetheric waves, could be invaluable. (It took twenty years for the term *radio waves* to come into use.)

Early on, experimenters realized that various frequencies of radio waves have various propagation characteristics. Short waves can reflect off the ionosphere, more than fifty miles overhead, and bounce back to Earth in distant places, but locally they cannot bend, or diffract, around obstacles, such as mountains and buildings. Hence they are mainly useful in local, line-of-sight situations, such as (these days) when airplanes need to communicate with nearby control towers. Long waves, by contrast, can diffract around mountains and buildings and can more easily follow the curvature of the earth. Thus scientists realized that the radio spectrum offered a range of possibilities and that various wavelengths could be used for specific purposes.

It took just a few short years after Hertz's 1888 discovery for entrepreneurs to start to recognize its technological potential. In 1892 the physicist William Crookes wrote about the possibilities of "wireless telegraphy" based on Hertzian waves. But a few other inventors, including Serbian American Nikola Tesla, disagreed, considering the waves useless for communication. Tesla mistakenly believed that this new form of invisible light, just like visible light, could not transmit farther than the observer's line of sight; meaning that if the target was not in view the waves wouldn't get there. (Tesla did initially experiment with radio waves, but never followed through because he was too preoccupied with attempts to send electricity through the air, which he regarded as a more promising and useful technology. Time proved him wrong on both counts.)

Inventors entering the business of wireless telegraphy soon were able to communicate across a range of several miles. Actually, in the United States, Tufts University professor Amos Dolbear had created a half-mile radio transmitter and receptor in 1882, before Hertz had officially discovered his waves. Unfortunately for Dolbear, he was too slow in applying for a patent and thus missed out on being credited with the invention of the radio. That honor went to the Italian Guglielmo Marconi, who in 1897, a mere nine years after the publication of Hertz's paper, founded the Wireless Telegraph and Signal Company in Britain, which soon became the Marconi Company.

By 1899, wireless telegraph communications were crossing the English Channel; ever-larger antennas, stronger signals, and longer distances were announced almost monthly. Radio receivers, which started out as mere metal poles, began to use a more active receiving process that involved *coherers*. These were tubes containing two electrodes spaced a small distance apart with metal filings in between. When a radio wave reached the coherer, the metal particles clung together, or cohered, which diminished the high resistance of the device and allowed an electric current to flow through it. Thus hit with a radio signal, the current could activate a Morse code paper recorder and cause it to make a record of the signal. For their time, these coherers were marvelously high-tech. The invention of vacuum tubes made detecting radio waves far easier, and greatly extended their range—all this before the start of World War I.

By 1911, many ships were equipped with wireless telegraphs, and their range spanned hundreds of miles. Marconi's device is what saved the lives of the *Titanic*'s seven hundred survivors when the sinking ship's SOS call was received by the *Carpathia,* some fifty miles away.

Carrying voices or music—audio—required a different technique. Instead of discrete individual sparks, which was all that was necessary to produce Morse code, continuous radio waves needed to be sent and received. Modulations, or changes, in that continuous signal could convey the rapid-fire information contained in the human voice and the notes of the musical scale.

The first voice was wirelessly transmitted on June 3, 1900, by the Brazilian priest Roberto Landell de Moura. His first public experiment, in front of journalists in São Paulo, Brazil, broadcast his voice a distance of five miles across that city. He ultimately applied for and received three US patents after he took his equipment to Washington, DC.

The United States then woke up with a vengeance to this new technological wonder. After Westinghouse engineers invented the vacuum tube detector, radio-wave reception became far sharper. The first radio program broadcast, from Ocean Bluff–Brant Rock, Massachusetts, to ships at sea, took place on Christmas Eve in 1906. Using his brand-new synchronous rotary-spark transmitter, American inventor Reginald Fessenden played "O Holy Night" on the violin, then read a passage from the Bible.

The technology for this first-ever event was the same that powers AM radio broadcasts to this day. You see, a continuous wave is nice, but information—whether voice or music—requires pulsations, so the signal has to vary. Early radio waves could be stopped and started, which was fine for sending the dots and dashes of Morse code. But how to make them continuously change?

There are two choices. Each wave has a height, essentially a strength or loudness. These waves are electrically converted from sound to electricity by a microphone. By rapidly making the electrical signals it receives louder and softer—i.e., their peaks taller or shorter—a transmitter coupled with the appropriate receiver can encode and detect these rapid fluctuations and translate them into the original sounds by using a rapidly vibrating magnetically driven speaker. This is *amplitude modulation,* or AM.

The other method is to rapidly change the frequency of the waves, meaning subtle variations in their wavelengths, by causing a sudden spurt of more or fewer waves. This *frequency modulation,* abbreviated as FM, allows for a greater clarity of transmission and reduces static and interference from electrical equipment. Early radio was exclusively AM; the first experimental FM broadcasts didn't begin until 1937, the process having been patented four years earlier by inventor Edwin H. Armstrong.

In June of 1912 Marconi opened the world's first radio factory. The first news program was broadcast on August 31, 1920, by station 8MK in Detroit, Michigan, which survives today as an all-news station affiliated with the CBS network. As for the first public entertainment broadcast in the United States, that, too, came in 1920, with a series of Thursday night concerts.

If only Hertz, Maxwell, Faraday, and the other early pioneers of wireless telegraphy and radio could see the world now. They would stand in awe of what would probably seem to them like magic. Consider the GPS system found in virtually every car and cell phone. How much do you know about the work it's doing constantly? Atomic clock signals broadcast by two dozen satellites send out time signals using short radio waves—nowadays categorized as microwaves. Your GPS receiver needs to receive only four of those signals to determine your exact location on the planet. It knows the right time to a billionth of a second and immediately discovers that all those satellite signals it's receiving are delivering the wrong time. Wrong because those Hertzian waves, traveling at light speed, require a little bit of time to reach you—about one-seventeenth of a second—from their orbital height of eleven thousand miles.

The delay is caused by the time it takes light to travel. Your receiver, perceiving a certain precise delay from satellite A, a different delay from satellite B, and so on, can calculate how far away each orbiter must be to produce that out-of-date time signal. Then it calculates the one spot on earth where you must be located in order for you to be positioned at that exact distance from each satellite. Voilà: your location, to an accuracy of a few feet. Because your GPS also knows where you are each time it performs a triangulation, it knows how fast you must be moving and in what direction. Equipped with onboard maps, it can then determine how long it will take you to reach Cleveland or any other place you're going.

All using radio waves.

This wouldn't work if light were so fast that it took no time at all to travel. And it wouldn't be practical if light's speed were too slow—matching that of sound, say. So if some of our major technologies revolve around the speed of radio waves and other forms of invisible light, maybe we'd better take a closer look at that speed.

CHAPTER 11

The Speed That Destroyed Space and Time

Light—visible and invisible—has a weirdness that extends far beyond its "not really there if no one is looking" quality. Many of its oddest effects involve its speed, probably the most famous velocity in all of science. You see, at high speeds or in places with a strong gravitational pull, all kinds of funny things happen, and since they affect light, they affect a lot of other things as well. At superhigh speeds, time slows down—not good if you have a boring job—and distances shrink. This phenomenon, known as time dilation, is amply explored in science fiction, as it was in the 2014 blockbuster movie *Interstellar,* in which astronauts on the surface of an alien planet experience the passage of only a few hours while their colleague in orbit grows several decades older. But the contraction of the distances between objects is much less widely appreciated.

The fact that neither time nor space has definable boundaries was first expressed in a famous equation devised in the final years of the nineteenth century by Hendrik Lorentz. A few years later, when Einstein explained them, he used the same math, which is why the specifics of the mutation of time and space is still called the Lorentz transformation. The equation shows some of the astounding possibilities: if you could travel at 99 percent of the speed of light, the universe would suddenly become seven times smaller than it is now. A star around fourteen light-years away (such as Altair, in the Summer Triangle) would magically float to a distance just two light-years away—reachable before you'd aged very much. At that speed, a living room shrinks to barely three feet wide.

As you zoom faster, the effect becomes truly dramatic. Traveling at 99.9999999 percent of the speed of light would produce a dilation factor of 22,361. Your rocket's clock would tick off just one year while more than 223 centuries *simultaneously* elapse back on earth. In addition, all distances would shrink by this same percentage. You could reach the core of our galaxy in a single year—as compared to four hundred million years using today's fastest rockets. With that much territory at your disposal, your social life could expand enormously. You could hold your next party near the black hole at the galaxy's center and toss popcorn in just for laughs.

But you'd be laughing alone. Back on earth, clocks would continue to tick and tock at their old customary rates. When you returned from your round-trip, the party would really be over. While you and your crew experienced two years and would look only that much older, fifty thousand years would have elapsed back on earth. The map of the world would have changed utterly; customs and language would be unrecognizable to you. There would even be evolutionary alterations in some life forms. You'd be seriously old-fashioned, and it would go far beyond whether bell-bottoms and wide ties had come in or gone out. You'd be lucky if they didn't throw you in a zoo.

When pondering the space- and time-altering effects of approaching the speed of light, the easiest way to keep things straight is to remember that, to you, time always passes normally and never seems to change. If you're lucky enough to live for eighty-five years, you'll experience that passage of time in the normal way, regardless of your velocity. It's just that others watching you through telescopes will age differently and observe you aging slowly, even though you experience no such alteration. Bottom line: it's always "the other guy" whose time passes weirdly. It's never you.

But you do experience shrinking space. When you approach the speed of light, the distance ahead to your destination does indeed contract, so you get there much sooner than you would if space weren't shrinking. In addition, at close to the speed of light, everything in the universe would seem to lie directly ahead, no matter which way you point your rocket.

Okay, this requires some explanation: it's the principle of *aberration*. When walking briskly through a rain shower, we must tilt our umbrella forward a bit to keep from getting wet. And when we drive through a snowstorm, the flakes seem to come from straight ahead while the rear window hardly gets hit at all. These are examples of aberration, which means a shift in the position of an object. Light, too, gets changed in this same way. Earth's motion around the sun at a mere 18.5 miles per second shifts the night's stars so that they're slightly displaced from where they'd appear to be if we weren't moving. If we traveled faster, this effect would increase until, at just below light speed, everything in the universe would seem to be located dead ahead. So looking out our rocket's windshield, we'd see a single dazzling "star," which would actually be everything in the universe clumped into one brilliant ball. Out the rear window we would see inky blackness, because space would be so warped that nothing at all would lie in that direction.

In short, a star's position is changed because of our own motion through space!

Light possesses a magician's chest full of illusions and tricks. Its speed, that famous constant of 186,282.4 miles per second (or 299,792,458 meters per second), refers to its velocity through empty space. But light moves more slowly through denser media such as glass (120,000 mps) and water (140,000 mps). So the light rays that convey all the colors and shapes outside your window through the pane and to your eyes abruptly slow down when they strike the glass. Then they instantly speed up again once they're finished penetrating it. Light travels slowest of all through a diamond, because each color contained in the light moves at its own distinct velocity through the gemstone. This difference in speed is why a diamond flashes, giving it its coveted brilliance.

Despite many attempts, nobody could figure out the speed of light until a few centuries ago. It was just too fast to measure. Not that many determined "natural philosophers" (as scientists were then called) didn't make heroic efforts to do so. In 1629, Dutch scientist Isaac Beeckman positioned large mirrors at various distances from gunpowder, which he'd explode, creating a flash. His helpers were instructed to observe the flash directly as well as watch its reflection in a distant mirror, light from which would thus need to travel a greater distance to their eyes. Would they detect any lag or delay between the two flashes — the direct event and its reflection in a faraway mirror? Answer: no.

Three years later, Galileo gave it a try. The cranky bearded genius stood on a hilltop and positioned an assistant with a shuttered lantern on another hilltop, one mile away from him. Galileo opened his lantern, and the assistant was supposed to hit the quick-release shutter to his own lantern as soon as he saw Galileo's light. Galileo would then measure how long it took before he saw the "responding" light from the other hilltop. By measuring the elapsed time and knowing the distance between the lanterns, he would determine the speed of the light.

In reality, light's round-trip travel time between two hilltops a mile apart would be 1/100,000 of a second. Good luck, Galileo. He concluded, "If not instantaneous, [light] is extraordinarily rapid." He ended up declaring that light travels at least ten times faster than sound. (Sound travels around one-fifth of a mile in a second, or around a million times more slowly than light.)

The Danish astronomer Ole Rømer finally obtained the first reasonable light-speed measurement without having to trudge up any hills. In 1675, when he was thirty-one years old, he explained why Jupiter's four bright giant moons all bewilderingly sped up in their orbits for half of each year, whenever Earth was heading in Jupiter's direction. It's logical, he explained: the *images* of Jupiter's moons' positions reach us sooner under those conditions because each second their light doesn't have quite as far to travel as it did before. The end result is to make them seem to zoom around that planet in high-speed Charlie Chaplin fashion. Not knowing the precise distance to Jupiter was his only weak link; nonetheless he correctly nailed light speed to within 25 percent accuracy.

The highest radiation levels in the solar system surround the giant planet Jupiter. Its four bright moons, oddly, seem to orbit faster for six months out of the year than they do during the other six months. Danish astronomer Ole Rømer used this curious fact to pin down the speed of light with reasonable accuracy in 1675. *(NASA/JPL)*

These days, to measure light, we use an apparatus first devised in the middle of the nineteenth century by the famed French scientist Léon Foucault. It works by shining a light beam at a rapidly rotating mirror, which reflects the beam to a distant fixed mirror, which then reflects it back. While the photons are making their little round-trip journey, the whirling mirror's face slightly changes its angle so that the final bounce gets reflected to a slightly different position on a graduated scale in an optical detecting device. Knowing the speed of the whirling mirror (Foucault's spun five hundred times a second) and the distances between all the mirrors, and taking into account the amount of light-path deflection read from the scale, one can nail light speed to an accuracy of one part in a million. I've done it myself; these days it is almost routine among scientists.

Light is indeed superfast, but we can now comprehend the speed; it's not infinite. If only the ancient Greeks could be here to share with us this small miracle—that we can actually measure the fastest thing in the universe.

We use light's steady speed in our technology and in our scientific experiments—and from it we learn new things. For example, several college research programs bounce laser flashes off the three corner cube reflectors left on the moon by the Apollo astronauts. The delay in receiving the reflection is always around two and a half seconds; precise measurements allow us to gauge the moon's changing distance from Earth to an accuracy of within one inch. It's revealed that the moon is constantly moving farther away from us at the rate of one and a half inches a year.

Let's saddle up and picture ourselves riding on a photon. In one second of zooming at light speed, we could circle our planet eight times. One hour at that speed would carry us to Jupiter. But reaching the nearest star, Proxima Centauri, would require that we remain in that saddle for 4.3 years. And, alas, the nearest spiral galaxy could be reached only if we continued at light speed for two and a half million years.

What about a much shorter commute? For example, zooming at light speed for just a thousandth of a second? That would let us commute from New York to Washington, DC. In one millionth of a second, we could cross three football fields. And in a billionth of a second we'd travel 11.75 inches—essentially a foot.

That's a fun statistic, because it means that everything

around you is seen the way it was as many nanoseconds, or billionths of a second, ago as it is feet away from you. You view a friend sitting across the room twenty feet away not as she is now but rather as she was twenty billionths of a second ago.

Since no image or information can exceed that velocity, we can never know what things are like *now* anywhere in the universe. Indeed we usually don't even try to know. Instead we define *now* as "whenever an image arrives at our eyes." We say, "Look how Jupiter and Saturn are passing one another in the night sky!" and no one bothers adding, "Or at least look at the way they were passing each other when the light we see now started our way an hour ago."

The gap, or discrepancy, between the way things appear to our eyes and what the current reality is grows ever greater as we look farther away, and it equals the age of the visible universe when we peer 13.8 billion light-years away. This, then, is the boundary of observable reality, beyond which nothing can ever be seen or known.

There's no way of getting around this limit. We view stars in the Whirlpool galaxy as they were thirty-five million years ago, and there is no possible method of obtaining news of their current state. Nor, watching us through some supertelescope at this moment, could Whirlpoolean extraterrestrials see anything but Earth thirty-three million years before the first appearance of humans. Probing us using invisible forms of light wouldn't change anything. Anyone monitoring our radio and TV signals, or the infrared radiation given off by the heat of our bodies, would confront the same limitation, because these waves, too, travel at light speed. For the same reason, no technique or clever design could alert us that a laser beam or radio signal from an extraterrestrial civilization was en route to us until it actually arrived.

Probably the most intriguing of all aspects of light speed is a photon's perception of totally frozen time. If you could walk a mile in its shoes (so to speak) you'd experience yourself as everywhere in the universe at once.

This happens because light travels in a fundamentally different way from the way we travel. We perceive ourselves as moving through both space and time. Actually, when we're sitting still, we don't move through space relative to objects around us, but we still must move through time at the rate of one day every twenty-four hours, even if we don't want to. We age. As we walk, we travel through space and continue moving through time. Here's the astonishing thing: as we go faster, we traverse more space, but our journey through time is *reduced*. Our time slows down from everyone else's perspective. At just below light speed we are cruising through a lot of space but barely moving through time at all. The more you travel through one facet of space-time, the less you travel through the other. You can't fully do both simultaneously. That's what Einstein figured out, although most of us still don't grasp the enormity of the concept a century later.

Light exists at the extreme end of this phenomenon. Its photons only move through space. They experience no time at all. Thus they cross the entire cosmos in zero time, which means that from their perspective, distance separations simply do not exist. If you aim a camera with a flashbulb out a window toward the sky, the moment you pop the flash the pulse of light has already arrived at the far end of the universe, *from its perspective*.

It's very strange and unintuitive stuff. And yet light always moves at its famous constant speed no matter who does the observing. Light occupies a sort of higher, surer reality than the things we once thought we could count on, such as the interval between the ticks of a clock.

No wonder, then, that one of the first passages in Western religious scripture is "Let there be light" and at least one Eastern religion speaks about ultimate reality as a "clear light." The scribes somehow sensed that light exists in a more secure realm than the mere spatiotemporal dimensions in which we spend our everyday lives.*

Someday we may figure out how to exploit the time-warping qualities of light speed. Imagine, if and when our propulsion abilities are up to the task, we could go anywhere in the cosmos while hardly aging at all.

The only problem with truly distant high-speed travel: the earth will be eons older when you return. Your descendants will have evolved: they will no longer be what we perceive as human. Your jokes won't get a laugh. Records of your departure will have been lost thousands of years earlier. Your language will be unintelligible.

It's a good news, bad news kind of situation. Without violating any of science's laws, you've not only seen impossibly distant realms but also lived to witness eons of Earth's history. On the other hand, you might have been wiser to have never made the U-turn.

---

* Light's speed as an immutable constant brings bewilderment when students confront the concept of redshift and blueshift. Although each photon of light hits you at precisely the same speed whether you're crashing into it head-on or racing away from it, your relative motion to light does cause its waves to either scrunch up or spread apart. Since all but a handful of galaxies are racing away from ours, the light from all their countless billions of stars is stretched out and redshifted, meaning that their light has been shifted out of the visible range entirely and into the invisible infrared range. We could only view them, as William Herschel so presciently suspected early in the nineteenth century, with an infrared detecting telescope as opposed to an optical one. And indeed, astronomers increasingly use infrared telescopes when studying galaxies.

CHAPTER 12

Microwaves Everywhere

Pop a bag of popcorn into the microwave and hit a button or two. When the "It's ready!" beeps begin, you know the waves have stopped even as the popcorny aroma spreads throughout the house. And maybe some questions are flashing through your mind as insistently as the beeping. Are these microwaves safe? What actually makes those kernels pop? When people think about invisible rays zapping their bodies 24-7, they think of microwaves. Some people even use the word *zap* to describe what a microwave oven does: your mac and cheese isn't hot enough? Just zap it in the microwave for a few minutes!

Microwaves were often lumped in the same category as the radio spectrum. The word *microwave* wasn't even coined until 1931. But in the years before World War II, major technological uses for microwaves materialized, opening the door for their immense popularity. It was only then that the shortest-waved section of the radio spectrum earned its own discrete designation.

Since they are now customarily treated as an entity unto themselves, let's invite them to join the party. Let's define radio waves as each having a maximum length of one hundred thousand miles and a minimum length of a foot. Microwaves begin at that shortest point, with a maximum one-foot distance from crest to crest and a vibration rate of one trillion waves per second. When we reach the shortest end of the microwave spectrum, we find its waves spaced an eighth of an inch apart. Any closer together, and waves would earn membership in the infrared club.

Early on, within a decade of Hertz's discovery of radio waves, researchers realized that whereas light waves diffract around obstacles, the shortest radio waves can easily focus on a target. Thus they are excellent at point-to-point communication. Unlike longer radio waves, which bend around objects, bounce off the ionosphere, and are good for widespread radio transmissions, the shortest waves are perfect for things like sending a signal from a plane to a control tower. They operate in a strictly line-of-sight manner. They also crisply bounce off metal objects.

German inventor Christian Hülsmeyer quickly exploited this capability. In 1903, he used what we now call microwaves to create the first-ever ship-detection apparatus, intended to help vessels avoid collisions in fog. This device, which he patented, reliably "saw" ships up to five miles away in zero-visibility conditions. Hülsmeyer could even determine the unseen ship's bearing (direction), thanks to his device's rotating parabolic receiving dish. But it could not determine range (meaning distance). Despite Hülsmeyer's obtaining financial backing and performing an initial successful demonstration for the Holland America Line, interest in his device inexplicably sputtered, and within a few years his company had to be dissolved.

However, others soon picked up the ball and ran with it. By World War I, several inventors had developed similar but improved systems that used triangulation to obtain both bearing and distance readings over short ranges. During the following decades, and particularly in the years just before World War II, thanks to increasingly intensive research conducted in Britain, short-frequency radio waves successfully detected not just ships but also incoming planes—and this microwave-utilizing technology was called radar, an acronym for "radio detecting and ranging."

The creation of systems capable of putting out short pulses of radio energy, and the use of oscilloscopes to monitor delay times between the outgoing pulse and receipt of the echo, were the keys to determining both the position and distance of ships and planes—the basis of the radar systems that saved Britain.

Radar should have saved more than 2,400 lives on December 7, 1941, but human judgment got in the way. One of the first US radar installations was completed atop Kahuku Point, on Oahu's northernmost tip. This facility detected the first wave of Japanese aircraft on their way to attack Pearl Harbor when the planes were still 132 miles to the west. That sleepy Sunday morning, the radar operator brought this seemingly impossible collection of hundreds of incoming planes to his superior's attention. But because the system had been in operation for only two weeks, and because such an aggregation of aircraft was so unusual and seemed so unlikely, the supervisor dismissed the countless images on the oscilloscope as a fluke, or perhaps a flock of birds. The system worked perfectly as designed, but the critical warning was ignored.

After the war, further improvements to microwave technology resulted in the invention of Doppler radar, which allows the operator to determine the speed of an object moving toward or away from the beam of radio waves. This principle is illustrated by a speeding ambulance. As the ambulance comes closer, you'll notice that the siren rises in pitch while the interval between each of its warbles is reduced. Then suddenly, the second after it passes you, those warbles seem stretched out and more widely spaced, and the pitch simultaneously lowers. The explanation for this effect, which applies to light as well, was first explained by the Austrian physicist Christian Doppler in 1842. It's simple: although light always travels at a constant speed, its waves nonetheless either compress or stretch out depending on whether the light or the observer is approaching or receding. The wavelength change happens in all directions except sideways or tangentially, so that a radar gun will not be able to read a baseball's speed if it's traveling sideways to the instrument—from right to left, say. In the case of visible light, the effect is to make the light of approaching objects appear blue, because its waves are being crammed together, or shortened, and because blue light has shorter waves than red light. This effect is known as a blueshift.

All but half a dozen of the universe's nearest galaxies are rushing away from our Milky Way, so they appear more red to us, the result of the famous redshift that occurs when a light source is flying away from us. Indeed, very distant galaxies, which rush away at a goodly fraction of light speed, appear more than merely reddened. Their images are shifted beyond the visible portion of the spectrum, and the galaxy only appears to us in the infrared part of the spectrum. This principle works just as well with light that was invisible from the get-go. When an object is approaching a radar antenna, the radio waves returning from it become increasingly compressed the faster it moves. Conversely, returning waves from objects moving away become increasingly elongated and have a longer wavelength and lower frequency. By measuring this frequency shift, you can pin down the speed of an object toward or away from the antenna.

These days, law enforcement uses Doppler radar to dispense

traffic tickets. Coaches use it to time athletes' running speeds and pitched baseballs, and meteorologists use it to study the detailed rain motions within a thunderstorm.

Synthetic aperture is yet another widely used modern type of radar. This involves the use of a *moving* radar device. We know that the larger the dish, the better the resolution, so one alternative to giant arrays of radar dishes is to mount a radar dish on a moving plane (say), which then keeps sending pulses and receiving the echoed signals from many "sending" and "receiving" locations. This in turn makes the device perform as if its dish were much bigger. This type of radar creates images of landmass formations and military targets in exquisitely fine detail, even if the objects are just a few inches across.

All these technological marvels really began with the magnetron tube, invented in the 1920s and continually improved through the early 1940s. The magnetron tube, which produces microwaves, is based on the same simple cathode-ray tube that fascinated the world in the latter half of the nineteenth century. In the twentieth century, until flat screens came along, it was an essential component of the picture tubes used in old-fashioned TVs. By introducing a powerful permanent magnet and cleverly spaced cavities into the tube, the electrons flying from the cathode are forced to change direction, which makes them emit microwaves. This shouldn't sound too implausible if we remember that the motion of electrons is what produces every type of light there is, except for gamma rays.

Beyond all the marvelous radar breakthroughs, the microwave property that most changed our lives was discovered by accident near the end of World War II. That's when an electronics genius named Percy Spencer enters our story.

He was born outside Boston in 1894, and his childhood wasn't easy. His father died when he was eighteen months old, and his mother soon left him to be raised by an uncle. His uncle died when Percy was seven years old, leaving him an orphan. Spencer subsequently left grammar school to earn money to support himself and his aunt. Between the ages of twelve and sixteen, he worked from sunrise to sunset at a spool mill. Then he discovered that a local paper mill was soon to begin using electricity, a concept little known in his rural area, and he accordingly began learning as much as possible about it. His self-education was so thorough that when he applied to work at the mill, he was one of three people hired to install electricity in the plant, despite never having received any formal training in electrical engineering—not to mention never having finished grammar school. At the age of eighteen, Spencer decided to join the US Navy. He had become interested in wireless communications after learning about the wireless operators aboard the *Titanic*. While with the navy, he made himself an expert on radio technology by reading books, even while standing watch at night.

A quarter century later, Spencer had, astonishingly, become one of the world's leading experts in radar-tube design thanks to his work with microwaves at a young defense company named Raytheon. As the chief of its power tube division, Spencer kept finding new ways to improve radar design and production. One of his breakthroughs enabled the company to increase its output of radar units from a dozen a day to 2,600 a day. His staff grew from fifteen employees to five thousand. By the end of World War II he had earned more than two thousand patents and was awarded the Defense Department's highest civilian honor, the Distinguished Public Service Award.

Soon after the war ended, while inspecting one of his Raytheon laboratories, Spencer paused in front of a working magnetron tube. Suddenly he felt something soft and sticky in his pocket. His chocolate–peanut butter candy bar had begun to melt. Others had noticed this phenomenon before, but Spencer decided to investigate it full tilt. He brought in some popcorn the next day, and sure enough, his magnetron tube made it pop.

It turns out that microwave energy, though it bounces off metal, is easily absorbed by water. Thus anything with moisture in it—and this includes virtually all foods—will heat up in the presence of these invisible rays.

The first microwave oven Spencer and his Raytheon company produced weighed a third of a ton and was the size of a refrigerator. It was expensive, too, costing thousands of dollars in an era when a new car could be bought for one-tenth that amount. Only commercial kitchens and cruise ships found a use for it. Nonetheless, Raytheon hoped for more widespread adoption as they started marketing the oven under the catchy name of Amana Radarange.

As magnetron tubes got smaller and cheaper, so did microwave ovens. Home units were introduced in 1965, though they were not widely purchased until the 1980s, when plummeting prices made them irresistible. A microwave oven cost $495 in 1968, yet a sleeker model with more bells and whistles would set you back just $191 (in inflation-adjusted dollars) in 1986. Today, more than a billion microwave ovens are in use around the world.

Some wonder if food cooked in them is safe. The answer is

an unambiguous yes. We know this for two reasons. First, in some places, including Japan, they've been in widespread home use for fully half a century. If any negative health effects were to arise, we would have seen them long ago; Japan actually has the highest longevity rate of any nation on the planet.

Second, if your food (or anything else) is hot, it simply means that its molecules have been sped up. During frying, grilling, baking, or any other cooking process, infrared radiation from a flame or an electric heating element produces the jiggling atoms. In the case of microwaves, exactly the same thing happens, but with two important and interesting differences.

Microwaves do penetrate the interior of food, but they still mostly heat from the outside in, whereas a frying pan vigorously heats the outside layers first before the heat works its way inward. The fact that microwaved food contains both hot and cool interior portions shows that it cannot simply work by an even process of convection from the exterior to the interior. Also, microwave ovens produce standing waves (persistent patterns) and repetitive microwave eddies within the oven, which are uneven. An internal "stirrer" tries to make the waves flow through the oven evenly, but this never works perfectly. The rotating tray helps, but nonetheless some parts of the food typically receive more waves than others, which results in uneven heating. Moreover, wet parts of the food heat more rapidly than dry parts. All these are culinary drawbacks, but none produces health risks, though they help explain why noted chefs rarely recommend using a microwave oven for your gourmet cuisine.

As for your own safety when using a microwave, just check out the screening that seems to be part of the glass in the door.

The size of the holes is no accident. They're designed to be smaller than the microwaves, which, as you may recall, vary from around half an inch to one foot across. None can fit through the holes, and therefore not a single microwave can leave the oven to heat the gum you're chewing as you stand there, impatiently waiting for the beeps. Bottom line: independent testing consistently shows that microwaves never escape from the oven. So yes, you can keep standing there, watching the hypnotic spinning of your frozen burrito.

Anyway, what on earth is better than a microwave when we want to salvage week-old frozen pizza slices or enjoy some quick popcorn? Indeed, on your next movie night, take a second to salute Percy Spencer with your butter-coated fingers. Spencer was eventually awarded an honorary doctorate from MIT—not bad for a guy who never completed sixth grade.

We've established that food cooked in microwaves is safe to eat, but still the sight of hot dogs sizzling in a microwave oven—of microwaves cooking flesh—can give one pause. If they can do that to store-bought frankfurters, what can they do to us? You could ask Siri on your cell phone, but then again, cell phones use radiation in the microwave spectrum. More than a billion people use cell phones every day. You probably have yours close at hand right now. What emissions is it sending your way, and what effect are they having on you?

Spencer would probably never have worried about it. After two marriages and three children, Spencer went to microwave heaven in 1970, at the age of seventy-six.

CHAPTER 13

The Man with the X-ray Vision

We probably utter the word *X-ray* far more than the words for other kinds of invisible light. We know about infrared light, but we usually refer to it as heat, as in our bathroom's heat lamp. Ultraviolet light, too, is seldom referred to, and then only as something to avoid—for example, when we apply a UV blocker at the beach. But when it comes to Wilhelm Conrad Röntgen's discovery, we ask for it by name: "My doctor sent me for an X-ray." Maybe it has something to do with that nerdy *X,* a letter that has never lost its attraction for sci-fi lovers and gadget freaks.

Wilhelm Conrad Röntgen was born on March 27, 1845, in Lennep, Germany. His family was neither poor nor fabulously wealthy. His parents manufactured and sold cloth. When he was three years old, his family moved to the Netherlands, where he went to boarding school. Written accounts of his childhood reveal that Wilhelm had a deep love of nature and of trekking in the forest, a pastime he never abandoned. He was also handy, which is perhaps why at age seventeen he entered a technical school in the Dutch city of Utrecht. The oddest story of this period, though irrelevant to his later accomplishments, is that he was thrown out of that school for drawing a caricature of one of the teachers. In fact, although he'd been caught with that drawing in his hand, he'd kept his mouth shut and refused to rat on the actual artist. The culprit was the student at the desk next to his. The expulsion cost him; when he applied to Utrecht University, in 1865, to study physics, he was denied admission. But he wasn't giving up. He learned that students could enter the polytechnic institute in Zurich if they could pass its tough entrance exam; he took it and breezed through it easily.

Röntgen thus began his studies of mechanical engineering, and in 1869 he earned his PhD at the University of Zurich. Five years later he was appointed as lecturer at the University of Strasbourg, and the following year he landed a job as professor in the agricultural academy at Hohenheim. In 1879 he accepted an invitation to become the chairman of the physics department at the University of Giessen, and—continuing his ascent up the academic ladder—accepted the same position in 1888 at the University of Würzburg. This was an important move, because his colleagues at that point included several others in our larger story, including radio-wave pioneer Heinrich Hertz as well as Hendrik Lorentz, who predicted the existence of the electron.

Röntgen met his future wife, Anna Bertha Ludwig, there, too, at a coffeehouse. A tall brunette and his senior by six years, she was the thirty-two-year-old daughter of the proprietor. They married the following year, 1872. Although they were unable to have children, in 1887 they adopted the six-year-old daughter of Anna's brother.

Röntgen worked hard and, starting in 1870, had several papers published in popular areas of research at the time, including the thermal conductivity of crystals and the electrical properties of quartz. But it is the strange new rays to which his name came to be forever linked that interest us, and this is what held his curiosity and inspired his hand-built experiments in 1895. Working without an assistant, he pursued a major late-nineteenth-century enigma: the strange invisible energies that materialize when a high-voltage electric current is sent through tubes containing so little gas that what they hold is nearly a vacuum. He was probing those mysterious "cathode rays" before they were found to be electricity.

On the night of November 8, 1895, he noticed an odd thing. He'd wrapped the glass vacuum tube in thick black cardboard to block all light and placed a paper plate coated with a barium compound six feet away. When he shut off all the lights in the room, the paper plate glowed. Intrigued, he went further—blocking the glass tube with substances of various thicknesses to see which permitted these odd unseen rays to still make the distant barium glow.

For weeks he barely slept as he continued performing experiments with this strange new invisible emission that seemed to be reaching out from the glass tube and affecting objects many feet away. Then, on December 22, he decided to try something novel. He held up a photographic plate a few feet from the tube while the high-voltage current was turned on. Then he took his wife's hand and held it motionless in front of the plate for several long seconds. As he developed the photograph, there appeared a ghostly image of the bones of her hand, including her wedding ring, with fainter gray impressions of the surrounding flesh.

When he excitedly showed her the picture, Anna screamed. A perfectly natural response, considering that until then, no one had ever seen the skeleton of a person who wasn't long dead. "I am seeing my own death!" she exclaimed in horror. He reassured her that there was no danger. But her reaction may have rattled him. He became one of the first and only researchers to use a lead apron and other screening tools while in his lab to block these unseen rays from penetrating his body. And although he died twenty-eight years later of intestinal cancer, it's widely believed to be a coincidence, because his cumulative exposure to X-rays is not thought to have been excessive enough to cause the illness.

The photograph of Anna's bones and the ring on her finger proved that some new ray of invisible light was involved in creating the image. Previously discovered forms of invisible electromagnetism did not behave this way: they did not penetrate solid objects such as flesh and paper only to be stopped by denser materials such as bone. These new rays were obviously highly penetrating and did not reflect easily, as did other forms of light. Since their properties were mysterious, he assigned to them the classic mathematical expression of the unknown—the letter *X*—though most people wound up calling them Röntgen rays and continued to do so for many years to come. Röntgen published a paper about his findings on December 28, 1895, entitled "On a New Kind of Rays" ("Über eine neue Art von Strahlen"). The press soon got hold of it, and word spread quickly. It was reproduced in the US journal *Science* on February 14, 1896.

For years, the *X* in X-rays remained appropriate, as their true nature continued to be enigmatic. It took until after the turn of the century for Max von Laue, who won the Nobel Prize in 1914, to show that the way X-rays diffract or interfere with each other when encountering crystals proves that yes, they're truly electromagnetic in nature, like all other forms of light. Moreover, they vibrate at astonishingly high frequencies—on the order of a million trillion waves per second. The distance between each of their waves is only a billionth of a meter, or 1/30,000,000 of an inch. In contrast with Hertz's newfound radio waves, which can be as long as a mile from crest to crest, the idea of waves no larger than atoms seemed preposterous—one reason why their true properties stayed hidden for the next half century.

Super-packed-together, super-high-frequency X-rays have powers and abilities that put them worlds away from, say, familiar red sunset light. Indeed, they opened the door to technologies that still seem straight out of science fiction. But before we explore these, let's close our story of their discoverer.

If the acclaim showered on Hertz for his radio-wave discovery seven years earlier was deafening, it was a mere whisper compared to Röntgen's instant celebrity. The potential to use Röntgen rays in medical diagnosis was recognized immediately. Numerous honors were lavished upon him. In several cities, streets were quickly named after him, and he received an avalanche of honorary memberships in learned societies around the world.

In spite of all this, Röntgen remained more than merely modest. Though he realized how enormously lucrative it would have been to patent his rays, he refused to do so, preferring to let the world freely use his discovery. And when he won the Nobel Prize in Physics, in 1901—the first ever—he donated his prize money to his university.

In 1914, Röntgen intended to immigrate to the United States, having been offered prestigious teaching and head-of-department positions at Columbia and other universities. He bought steamship tickets for his family. The sudden outbreak of World War I put his plans on hold. It was a delay that, sadly, proved to be permanent. After the war, the runaway inflation then affecting the Weimar Republic virtually bankrupted Röntgen, though he managed to hang on to his modest summer house in Weilheim, in the foothills of the Bavarian Alps. To the end, he never lost his lifelong love of nature and only stopped his routine of trekking through the Alps when his intestinal cancer became too advanced. At the age of seventy-seven, four years after his beloved Anna passed away, he joined her.

CHAPTER 14

Röntgen Rays for Everyone

The ink on Wilhelm Röntgen's paper announcing his discovery, published on December 28, 1895, had barely dried, and already it was causing a sensation around the world, one that would continue through the following year. By the end of 1896, scientists had created thirty-two distinct X-ray tube models. More than a thousand published scientific papers on the new rays appeared during that single year. In the midst of this X-ray mania, no one suspected that this marvel of science might have a dark side.

Some science luminaries did at first dismiss the discovery out of hand. Lord Kelvin, who had won international acclaim for successfully overseeing the laying of the first transatlantic telegraph cable, called X-rays a hoax and declared, very simply, that they did not exist. For a few weeks, many others in the sciences reacted with caution, too. In late January of 1896, the journal *Science* wrote with skepticism, "It is *claimed* that Dr. Röntgen has found the ultra violet rays from a Crookes vacuum tube penetrate wood and other organic substances, whereas metal, bones, etcetera are opaque to them."

But all skepticism disappeared by February of 1896. Even before the start of spring, *Scientific American* predicted that

Röntgen would be "immortalized for his discovery, and the year 1896 distinguished as the Röntgen Photography Year."

Thomas Edison, America's most celebrated inventor, wasted no time exploiting this new technology. It was still winter when he started building his own X-ray machine. In a letter, he told a friend that he wanted to perfect the X-ray technology "before others get their second wind." A mere twelve weeks after Röntgen's paper appeared, Edison had invented the fluoroscope, which, instead of taking a photograph, produced a sharp real-time moving X-ray image on a screen. Moreover, he announced, he would not patent this device but donate it freely to the world, as Röntgen had done with the X-rays it employed.

Following the strategy he used for his other vox-pop inventions, including the phonograph, Edison also created a version of the device—in this case, a portable fluoroscope—whose purpose was neither research nor diagnosis but rather entertainment. On May 4, in a New York City exposition hall, he unveiled the device and invited members of the audience to see their own bones on a glowing bluish screen. Crowds jostled to get close to the machine. Several hundred people at a time were allowed to file into the darkened room, taking turns, in pairs, putting their hands behind the screen. Demonstrating his showmanship, Edison arranged for a deep foghorn to sound each time the operator activated the device and ghostly images of the visitors' bones materialized. The deafening blast heightened the sense of drama. The crowds went nuts.

By that first summer, newspapers were filled with speculations and claims about the supposed health-restoring power of these new rays. The stories found an eager, uncritical audience: various "invisible rays" had been touted for years for their purported ability to rejuvenate the body. Electricity in particular was widely used as a tonic, supposedly offering cures for every imaginable ailment. In the mid-nineteenth century, a typical visit to a doctor included "electrotherapy" (or at least a conversation about it). Practitioners—some actual physicians, but mostly charlatans—moistened the patients' skin, attached electrodes to whatever site needed treatment, connected the cables to a powerful battery, and applied various amounts of "restorative" voltage. Treatment regimens were prescribed for menstruation problems, insomnia, anxiety, and a growing list of diseases and their symptoms. Depending on the complaint, the electrodes were placed in the rectum, the vagina, atop the base of the skull, on the uterus—pretty much anywhere. There were even "electrical baths" in which the patient experienced a whole-body tingling sensation delivered by wires suspended in warm electrified salt water. With electrotherapy then in vogue, it is small wonder that any freshly discovered rays would likewise be hailed as salutary—invisible spirits that can help us rather than phantoms that can scare us.

Quite suddenly, X-rays were all the rage. Articles about their powers appeared daily, many of them contradictory. X-rays could kill germs or they could restore friendly body essences. They could remove unwanted hair or stimulate hair growth. They could even restore sight to the blind.

This last claim had a strange etiology: many people claimed to be able to see X-rays that were aimed at their eyes. Today, few such vision-affecting claims are taken seriously—though nowadays it is of course unethical for any kind of X-ray beam to be aimed at a person's eyes. Nonetheless it's now generally acknowledged that humans can indeed perceive X-rays, at least sometimes, as a blue-gray glow. The mechanism by which we do this is unknown, and the whole issue is still bathed in mystery. The three most plausible theories posit that either (1) perception of X-rays results from the excitation of the retina's rhodopsin molecules, (2) X-rays directly excite retinal nerve cells, or (3) the observer experiences some secondary form of visual stimulation, such as X-ray induction of phosphorescence within the eyeball.

The first X-ray photograph after Röntgen's own was probably created a mere fourteen days after the discovery's publication. That's when Friedrich Otto Walkhoff took the first ever dental X-ray image, of his own teeth. He held a photographic glass plate in his mouth between his teeth and tongue and lay as still as possible on the floor for twenty-five minutes, with the high-voltage glass tube precisely aimed at his jaw.

Virtually no one called the new images of people's bones X-rays. They were shadowgraphs, cathodographs, electric shadows, and, most commonly, Röntgen photographs. Whatever you called them, they seemed too good to be true. By March of 1896, reports were circulating among physicians that X-rays had been used to find a bullet in the brain of a twelve-year-old child and to photograph a broken hip joint. People began building and selling homemade "Crookes tubes" at a furious rate: purchasers scooped them up as if they were today's latest iPhone model. The use of "Röntgen radiographs" (yet another early name for X-rays) skyrocketed. By the end of 1896 a Chicago electrical engineer named Wolfram C. Fuchs had performed more than 1,400 X-ray examinations, and doctors had started routinely referring their patients to "Röntgen specialists," who used primitive, often home-built machines that typically required an hour of nonstop exposure.

But for all their purported benefits, X-rays had some negative effects, too. It didn't take long for these to become obvious, because few practitioners bothered using lead aprons on themselves or the parts of their patients that didn't need to get zapped. Moreover, unfocused X-ray beams often penetrated walls, irradiating people in other rooms. Machine operators frequently tested their equipment by placing their hands in the beam. Many did this every single day. Wild overexposure was the norm. But while the world continued its celebration of the new invisible light, some in the medical profession were noticing odd, disturbing reports even during that first year. Most of those reports centered on skin blisters, which are now known to be a telltale sign that a person has absorbed a dose of at least 1,500 rads, a huge amount of radiation that greatly exceeds what most Hiroshima survivors endured in 1945. Writing in New York's *Medical Record,* Dr. D. W. Gage, of McCook, Nebraska, also cited cases of hair loss, reddened skin, skin sloughing off, and lesions. "I wish to suggest that more be understood regarding the action of the x rays before the general practitioner adopts them in his daily work," Gage warned.

That summer, at Vanderbilt University, a physician decided to experiment on himself before using X-rays to locate a bullet in a child who'd been accidentally shot. He placed the X-ray tube half an inch from his own head and turned on the tube for an hour. At first nothing seemed amiss. Yet three weeks later, all the hair in the directly exposed area fell out, creating a bald spot two inches in diameter.

On August 12, 1896, *Electrical Review* reported that one Dr. H. D. Hawks had given a demonstration with a powerful new X-ray machine. After four days he noticed a drying of the skin, which he ignored. His right hand began to swell and gave the appearance of a deep skin burn. After two weeks the skin came off the hand, the knuckles become very sore, fingernail growth stopped, and the hair on the area exposed to the X-rays fell out. His eyes were bloodshot, and his vision became considerably impaired. His chest was also burned. Dr. Hawks's physician treated this as a case of dermatitis. Hawks tried protecting his hands with petroleum jelly, then gloves, and finally by covering it with tinfoil. Within six weeks Hawks was partially recovered and was making light of his injuries. *Electrical Review* concluded by asking to hear from any of its readers who had had similar experiences.

September brought a more troublesome account. A man named William Levy had been shot in the head by an escaping bank robber ten years previously. The bullet had entered his skull just above the left ear and presumably proceeded toward the back of the head. Having heard about X-rays, he decided he wanted the bullet located and extracted. Levy approached professor Frederick S. Jones of the Physical Laboratory at the University of Minnesota. Cautious by nature, Professor Jones warned Levy against the exposure, but Levy was undeterred, and an X-ray was taken on July 8, 1896. The X-ray tube was placed over his forehead, in front of his open mouth, and behind his right ear. Levy sat through the exposures from eight o'clock in the morning until ten o'clock at night. Within twenty-four hours his entire head was blistered, and within a few days his lips were badly swollen, cracked, and bleeding. His right ear had doubled in size, and the hair on his right side had entirely fallen out. Professor Jones concluded that the one feature that was satisfactory to the patient was that a good picture of the bullet was obtained, showing it to be about an inch beneath the skull under the occipital protuberance.

Even more horrific were the side effects experienced by a glassblower at Thomas Edison's Menlo Park laboratory, Clarence Madison Dally, who insisted on certifying each Crookes tube he'd created. He tested the tubes' radiation output by placing his hands directly in the beam, turned to full power. Over the course of several months in 1896, Dally severely burned his hands—yet he continued that practice for another two years. In 1902 his right arm was amputated at the shoulder to arrest the spread of skin cancer, and two years later his left arm was amputated for the same reason. Clarence Dally died in October of 1904 at the age of thirty-nine—probably the first casualty of X-rays. His death horrified Thomas Edison and prompted him to discontinue X-ray research in his laboratory. Indeed, Edison was so appalled that he became afraid of X-rays. Later, when a dentist insisted on taking an X-ray to find the source of a persistent toothache, Edison asked the dentist to pull the tooth outright rather than subject him to the diagnostic rays.

By late in 1896, despite these troubling reports, the consensus among physicians was that X-rays were safe and that in cases of adverse effects the equipment had been improperly used. No one guessed that the deadliest X-ray perils take years to surface. For example, Friedrich Otto Walkhoff and Fritz Giesel established the world's first dental X-ray—or roentgenological—laboratory in 1896 and provided practitioners with images of the jaw and head for years to come. It took more than thirty years—until 1927—for Fritz Giesel to die of metastatic carcinoma, presumably caused by heavy radiation exposure to his hands.

So it was that those early warnings went unheeded. And to be fair to the profession, many medical reports were reassuring. For example, in Boston in 1897, a Dr. Williams reported that he had examined approximately 250 patients with X-rays and had not seen any harmful effects at all. In 1902, in a major Philadelphia medical journal, Dr. E. A. Codman conscientiously reviewed all papers on X-ray injuries. Of the eighty-eight X-ray injuries reported, fifty-five had occurred in 1896, twelve in 1897, six in 1898, nine in 1899, three in 1900, and one in 1901. Thus it seemed as if the practice of radiology was being steadily perfected and the risk dramatically declining. In actuality the decline may have simply been caused by the fact that neither X-rays nor the injuries they caused were novel or newsworthy as time went on and that therefore they went unreported.

It took more than a quarter century, during which time so many people had fallen victim to the misuse of X-rays that the dangers simply could no longer be ignored, before a book finally appeared to sound the alarm. *American Martyrs to Science Through the Roentgen Rays* was written in 1930 by the Boston radiologist Dr. Percy Brown, who himself died of cancer twenty years later, presumably from overexposure to X-rays.

Since then, the news has been a mixed bag. Radiology itself has become more cautious; these days it would be rare for a radiologist to be present in the room while X-rays are taken. On the other hand, CT scans, which deliver far more radiation than simple X-rays, have proliferated and have earned cautionary warnings from medical professionals who fear that their overuse is creating its own crop of future cancers.

But back in 1896, though no one suspected it, an even deadlier form of invisible energy was about to be discovered. This proved to be the ultimate: the most powerful kind of unseen light as well as the most mysterious. It provided the greatest amount of deadly radiation ever known, causing more than one hundred thousand deaths in a single week less than half a century later.

But before we open that story, we must look at exactly what radiation is and how it is quantified. In our modern times, this vital information is widely misunderstood.

CHAPTER 15

What's in Your Basement?

Did you know that a single whole-body CT scan often delivers more radiation than was received by Hiroshima survivors a mile from ground zero? Or that living across the street from a nuclear power plant for a full year gives you less radiation than eating a single banana? (That fruit contains a tiny bit of radioactive potassium-40, the main source of radioactivity in our bodies. It has a half-life of 1.42 billion years, so you might as well learn to like it.)

What is radiation? And how much of it is too much? For the vast majority of us, this issue is utterly bewildering. In our exploration of unseen rays and invisible hazards, we need a serious "time-out" to understand radiation.

Few terms are more misunderstood. Since the middle of the nineteenth century and the work of Faraday and Maxwell, all forms of light have been termed *electromagnetic radiation.* By this definition, a candle emits radiation, as do a night-light and the moon. Of course, such radiation is totally harmless.

By the final years of the nineteenth century, physicists began to discover invisible emanations in their labs and, as it turned out, throughout nature. Radium darkened photographic paper just the way X-rays did. This, too, was radiation, but an unknown kind. Was radium emitting tiny particles smaller than atoms? Or, instead, was it emitting unknown varieties of light rays? Whatever its nature, scientists wondered whether it was harmful or benign. Could it even be salutary? No one knew.

Soon all invisible emissions—whether particles or rays—that proved capable of affecting the body were labeled *radiation,* even though everyone knew that the word would also continue to be applied to harmless emissions such as those from starlight and fireflies. In short, the word *radiation* came to have at least two different meanings—which, confusingly enough, is still the case today.

So in this chapter, from this point on, unless I say "electromagnetic radiation," which just means some form of light, visible or invisible, *my* use of the word *radiation* will follow the most common usage and refer only to potentially dangerous emissions. Radiation can take the form of a submicroscopically tiny, high-speed, bulletlike piece of an atom, such as a proton, or a bit of light whose waves are so short that they can damage atoms they hit and thus induce cellular changes in living organisms.

Yet even that isn't the end of it. Electricity can kill you, yet nobody thinks of electricity as a form of radiation. No, to qualify as radiation, an emission must have the ability to fly through space or the atmosphere and not merely be transmissible through wires and such. Got all that?

Let's try this again: for the rest of this chapter, I'll call radiation anything invisibly tiny that doesn't require another substance to travel through but instead flies at superhigh speeds from point A to point B and can penetrate living tissue to affect animals and people. In discussions of biological damage, the word *radiation* means particles or energies that can alter atoms and therefore a cell's genes, causing birth defects and cancer.

* * *

As we've seen, radiation can mean tiny solid particles or it can constitute waves. Long waves can't damage atoms. That's why visible light, radio waves, and even microwaves cannot possibly injure genes and cause cancer. Living next to a cell-phone tower can jiggle entire atoms in your body so that they heat up tissue, but it can't actually break those atoms or cause tumors to form. (It still might not be good for you! And we'll return to microwaves in a later chapter.)

By contrast, gamma rays' and X-rays' short waves *do* break apart atoms, and UV radiation can, too. This is ionizing radiation, the bad kind, which can sabotage genes. Heavy, fast-moving subatomic particles such as neutrons can also destroy atoms, so they're often called radiation even though they're particles and not energy waves. (The distinction is blurry anyway, since all matter has a wavelike aspect.)

The cause of radiation can be as simple as an atom's electron falling inward, closer to its center, which creates a bit of energy that then flies off at high speed. Another common cause is something that befalls any atom with a large, unstable nucleus. Suddenly a piece of that nucleus can break away and fly like shrapnel through anyone and anything in the vicinity. Or that large nucleus, as it abruptly releases a fragment of itself, can emit a flash of energy, such as a gamma ray.

There is no way to predict exactly when any of these things will happen to a particular atom. However, a specific *kind* of nucleus—say, carbon-14, with its six protons and eight neutrons—always has a tendency to disintegrate within a particular period of time. Depending on the atom in question, that can be a short period, meaning a fraction of a second, or it can be years, even billions of years. An atom of the most common kind of uranium, for example, is most likely to "break up" into two smaller objects after 4.5 billion years. But if you're studying uranium atoms, it might be a frustrating wait, because the event might happen one second from now or *more than* 4.5 billion years from now.

So science considers only large groups of such unstable atoms and specifies how long we expect it will take for half the sample to disintegrate. In short, the issue is a statistical one. The half-life of uranium, 4.5 billion years, means that after 4.5 billion years we expect that half the batch of uranium we're studying will have come apart and changed into a different element (in this case, lead), accompanied by the release of particles or energy.

What's weird is that an atom has no "memory" of its past. It does not "age" with the passage of time. The chances of its breaking down remain the same throughout its existence, though again, each radioactive element's nucleus has its own specific decay-probability rate.

Half-Life of Selected Substances

Oxygen-15 122.24 seconds
Any neutron (outside an atom) 10.3 minutes
Carbon-11 20.334 minutes
Iodine-131 8.02 days
Sodium-22 2.602 years
Plutonium-238 87.7 years
Carbon-14 5,730 years
Uranium-238 4.468 billion years

Beyond figuring out the half-life of various radioactive substances, another challenge that confronted scientists in the late nineteenth and early twentieth centuries was to determine what constitutes a harmful *quantity* of radiation. Early radiologists exposed themselves to immense amounts of it to figure out where the safety boundaries lay. It was brave and ultimately killed many of them, although often not for thirty years or more after first exposure.

We've already seen how the earliest years of, say, X-rays brought disconcerting reports, but no widespread, consistent, hard-enough evidence surfaced to make radiologists take serious precautions. It wasn't until the 1930s that radiation's hazards started to become sufficiently clear—an awareness that became especially keen in the late 1940s.

In 1946 a statistical study of obituaries conducted by Dr. Helmuth Ulrich, published in the *New England Journal of Medicine,* found that the leukemia rate among radiologists was eight times that of other doctors. In 1956 the National Academy of Sciences supported those findings in a report that concluded that radiologists lived 5.2 fewer years than other MDs. In 1963, a study carried out by Dr. Edward B. Lewis found a significant incidence of deaths from leukemia, multiple myeloma, and aplastic anemia among radiologists, and two years later two Johns Hopkins researchers discovered that there was a 70 percent higher incidence of cardiovascular disease and certain cancers—and 730 percent more leukemia deaths—among radiologists than in the general population.

But back in the 1930s and even into the 1940s, the danger was still largely unknown. At that time, teenagers—including my own mother—were often treated for acne, pimples, and other routine adolescent rashes with massive X-ray beams, which would indeed dry out the skin and seemingly cure the condition. Until the mid-1950s, many shoe stores had customer-operated X-ray machines into which people inserted their feet. Then they could gaze as long as they wished to at the fluoroscope screen, pondering their bones and judging how their shoes fit. A friend of mine, in his mid-eighties at the time of this writing, witnessed seventeen H-bomb explosions while working as a technician with the navy. Wearing just casual clothing, he'd remove his protective glasses and gaze in awe at the mushroom cloud that he says was never more than ten miles away. That was in 1957, at Eniwetok, in the Pacific near the Mashall Islands. More than sixty years later, he's still healthy and active. Obviously the danger of excess radiation has a hit-or-miss component as far as genetic damage is concerned. True, the quantity of radiation matters mightily. But when you talk to friends and family about radiation, most act as if any and all forms of it are highly perilous. This widespread fear, bordering on phobia, reveals the public's deep ignorance of the subject.

Consider the Three Mile Island nuclear power plant accident of March 28, 1979. This worst-ever nuclear accident in the United States started with a mechanical fault, a stuck valve, and was exacerbated by an operator who misread a warning light. It all eventually led to a partial meltdown of the nuclear fuel and the creation of a hydrogen bubble that had the potential to explode and perhaps breach the containment building, which would have released significant radiation over a wide area. As it turned out, however, the total radiation released by the accident, as experienced by the two million people closest to the Pennsylvania facility, was 1.4 millirems, or 0.014 millisieverts. (Millirems and millisieverts, abbreviated as mrem and mSv, are the units we use to measure radiation.) The final report compared this with the 80 millirems per year that residents of Denver receive from living in that high-altitude city. As further comparison, a patient receives more than twice the accident's radiation, or 3.2 millirems, from a chest X-ray.

The Three Mile Island accident resulted in no private property damage and not a single injury to anyone. But in several almanacs and handbooks of industrial accidents, that event is categorized as a "calamity" or a "catastrophe." As far as the media is concerned, radiation accidents belong to an ultraperilous, headline-deserving class that bears no relation to actual injury or damage.

Perhaps even more surprisingly, experts from eighteen countries, when they assessed the Fukushima nuclear "disaster" of 2011, found that not only were there no fatalities, the likeliest number of future cancer deaths from the leaked radiation was also zero. Nuclear power plant "disasters" are thus far less perilous to health than is commonly perceived.

And yet at the same time, radiation *is* hazardous if you receive enough of it. Whereas some 23 percent of us will ultimately die of cancer, the rate is fully 1 percent higher for career pilots and flight crew members, thanks to the additional radiation they receive from routinely being up at high altitudes, where the cosmic-ray intensity is greater than it is on the surface of the earth. Every day, real lives are lost from radiation.

These days, fortunately, almost no one gets exposed to radiation in lethal doses. But it *can* be fatal. With an exposure of 700 rems (or 700,000 millirems, or 7,000 millisieverts), most people will die within a few days after very unpleasant initial effects, such as nausea, weakness, fever, and hair loss. Barring circumstances like the Chernobyl accident, which happened at a built-on-the-cheap reactor that lacked even a containment building, such intense exposures are unlikely for anyone on our planet except those in hospital settings, where a handful of people have indeed been fatally overexposed. These unfortunates were receiving routine radiation treatment when software glitches upped the dose to lethal levels.

One of these involved the Therac-25, a radiation-therapy machine first sold in 1982 that was produced by a company called Atomic Energy of Canada. Michael Mah, managing partner of QSM Associates, a software consultancy, told me that IT engineers still cite the Therac-25 as an example of what can go horribly wrong when safety depends solely on software.

The device was designed to treat cancer patients with beams of radiation. It could aim a beam of electrons for low-dose therapies if the tumor was not very deep, or it could be switched to an X-ray beam for deep or high-dose radiation treatment. Typically, a focused dose of around 70 rads, equal to around 70,000 millirems, was used. But in several instances, flaws in the software caused the patient to receive *one hundred times* more radiation than necessary—7,000 rads, or around 7 million millirems, which is a fatal dose. Between 1985 and 1987, the machine injured six people, three of whom died. One immediately felt the ultrahigh-dose electron beam as "an intense electrical shock," and he leaped up from the table. Screaming, he ran to the door and tried to escape.

Two previous models of the machine had hardware locks that prevented the device from accidentally switching to its high-dose setting. But the Therac-25 relied solely on software for protection against mishaps.

Because newly produced electronic devices typically have one error per five hundred lines of code, and because the Therac-25 had 101,000 lines of code, errors should have been anticipated.

As it turned out, a latent bug on a single line of code upped the radiation intensity. It was supposed to order the machine to increase the beam strength by a factor of six. Instead it commanded it to increase the radiation by $10E^6$, an *exponential* increase of six, which meant a millionfold increase. In practice, the machine couldn't deliver such power, so it responded by "maxing out" its radiation to full power. This "merely" boosted the dose by a factor of one hundred.

At the medical facility, the operator saw the message "malfunction 54" appear on the screen. However, the machine's instruction manual contained no information about what this might mean. So the operator simply overrode the error message by typing the letter *p*, for "proceed." The company's initial insistence that nothing could possibly go wrong kept the machines in place until new hardware protections and direct-dosage monitors eventually solved the problem.

The Therac-25 aside, fatal radiation encounters outside the realm of medical treatment have been extremely rare. But one unfolded in 1946. Louis Alexander Slotin, born in 1910, was a Canadian physicist who worked on the Manhattan Project at the famous (but at the time still secret) Los Alamos National Laboratory. Slotin's job and expertise, in the years just before and after the production of the first atomic bombs, involved experimenting with enriched uranium and plutonium to determine their critical-mass values—and then actually assembling them as weapons. He became known as the "chief armorer of the United States."

Slotin's hazardous "criticality testing" involved bringing sub-explosive quantities of those fissile materials to just below the critical-mass value, at which point a runaway atomic explosion would ensue. Imagine if flirting with the possibility of a nuclear chain reaction was part of your job every day. Scientists referred to this kind of work as "tickling the dragon's tail." It was Slotin, on July 16, 1945, who assembled the core for the device used in the first detonation of a nuclear weapon, held at Alamogordo, New Mexico, in the now famous test code-named Trinity.

Because he had done it several times before, perhaps Slotin was getting just a bit too casual. On May 21, 1946, with seven other Los Alamos experts standing off to the side observing him, Slotin placed two half spheres of beryllium around a fourteen-pound plutonium core. Slotin was holding the upper nine-inch beryllium hemisphere as though it were a bowling ball, with his left thumb stuck through a hole drilled in it, all the while keeping the half spheres separated with the blade of his screwdriver. It was critical to keep the hemispheres from approaching each other lest they trigger a nuclear reaction. Normally spacers or shims would have been used, but just as professional electricians often assemble live wires that you or I would never go near without first killing the circuit breaker, Slotin was performing with the mien of someone very comfortable with his job.

At 3:20 p.m., the screwdriver slipped, and the upper beryllium hemisphere fell downward. It fell less than an inch, and it only slipped for a single second—but an immediate "critical reaction" ensued. The sudden flood of radiation made the air glow a blue color—later recognized as the Cherenkov effect, caused by subatomic particles moving faster than the speed of light through air.

Simultaneously, everyone in the room felt a wave of great heat. Slotin later explained that he'd also tasted an intense sourness in his mouth. He immediately yanked his hand upward and threw the beryllium hemisphere onto the floor, which abruptly ended the chain reaction. But it was far too late. He must have already suspected that he'd received a fatal radiation dose. Indeed, he started vomiting before he even got to the hospital, and despite nonstop intensive care, including intravenous fluids, he died nine days later. His symptoms revealed what happens to any animal body when its cells are destroyed by radiation. His sad ordeal included swollen hands, severe diarrhea, intestinal paralysis, gangrene, and finally the failure of vital organs.

Of the seven others in the room, the person standing closest to the plutonium also needed hospitalization for almost a month, but he survived, although the accident left him with permanent neurological and vision problems. He died twenty years later, at the age of fifty-four. Another of the observers also had his life cut short and died nineteen years later, at age forty-two, of acute myeloid leukemia—a typical consequence of very high radiation exposure.

Such exposure to lethal radiation has, thankfully, been very uncommon except among Hiroshima and Nagasaki survivors. As much as they suffered, the vast majority of these survivors were exposed to far less than 300 rems, or 3,000 millisieverts, of radiation, a level that's fatal to half the people exposed to it.* Yet these were exactly the estimated exposures given to some patients who were subjected to early X-rays. Even when exposure reaches the level of 100–200 rems, or 1,000–2,000 millisieverts, people initially live through the experience, but their cancer risk is greatly increased.

And these were the levels that physicists started self-delivering during the early years of the twentieth century. The reason was simple: for more than two decades before that, there were as many

* To learn how radiation affected the health of Hiroshima and Nakasaki survivors, see this report: http://www.nytimes.com/gwire/2011/04/11/11greenwire-hiroshima-and-nagasaki-cast-long-shadows-over-99849.html?pagewanted=all.

"experts" who thought radiation was salutary as there were scientists who believed it might be harmful. Some early medical applications of radiation were in use for decades after the first nuclear weapons were developed. In 1981, two Montana "health spas" distributed pamphlets advertising the benefits of radon gas in curing "arthritis, sinusitis, migraine, eczema, asthma, hay fever, psoriasis, allergies, diabetes, and other ailments." The ads explained that sitting in abandoned mine shafts (after paying an entrance fee, of course) and breathing radioactive gases causes your joints to loosen and alleviates various aches and pains. The pamphlets failed to mention that at that time, the medical community had known for more than ten years that radon gas in uranium mines was causing miners to suffer a 500 percent increase in their risk of lung cancer.

Where is radiation found? Absolutely everywhere. It comes up from the ground and rains down from the sun and stars. Our atmosphere blocks some of it, but the higher up you go, the more you get. The average person at sea level gets 360 millirems (or 3.6 millisieverts), per year, of which 82 percent comes from natural sources. But thanks to the exploding use of CT scans during the past quarter century, some authorities now say that that the true average US radiation dose is more like 600 mrem annually.

Natural radiation is responsible for some of the spontaneous tumors that have always plagued the human race. But there is ongoing scientific debate over whether very low doses can harm us. Tibetans and Peruvians who live at high altitudes and therefore receive much more radiation than those living at low altitudes do not suffer from higher rates of leukemia, and a major 2006 French study showed no increased cancer incidence in children who live near nuclear power plants. There is experimental evidence from animal studies showing that exposure to radiation can cause genetic defects. However, studies of the survivors of Hiroshima and Nagasaki give no indication that this is the case in humans. Perhaps surprisingly, considering all the knowledge collected on radiation effects, there is still no definite consensus as to whether low levels of ongoing exposure to natural background radiation carries any health risk—even though risk has been demonstrated for exposure at a level just a few times higher. To be specific, consider the natural annual background rate of 360 mrem, meaning the amount one normally gets before receiving radiation from artificial sources, such as medical X-rays. This is widely considered to be harmless. But a single CT scan can deliver twice that much radiation, and medical authorities now assess a person's future cancer risk from a single CT scan as one in two thousand, which is certainly not zero.

But anyone concerned about radiation can easily calculate his or her personal annual dose. Start with the biggest sources:

Award yourself 26 mrem just for living on the surface of the earth. And don't blame Gaia—there is no planet in the entire solar system that gets less radiation than we do. We're actually a relatively safe haven.

Add 5 mrem for each thousand-foot elevation of your home. If you live in Denver you have to add 25–30 mrem. If your home is in Leadville, Colorado, up at around ten thousand feet, you receive 60 mrem more than the folks in Boston, at sea level, do.

Is your home stone, brick, or concrete? Add 7 mrem. These materials are naturally slightly radioactive. Only wood-frame homes are essentially free of radiation. Your real-estate agent never mentioned that, did she? Do you have a below-grade basement, which usually means a cellar without any windows or a cellar with narrow windows placed high up, near the basement's ceiling? If so, you have a huge potential danger if radon is present. The likelihood of this varies greatly depending on which area of the country you live in. For example, Northern California and northern New York State have virtually none. But most of southern New York State has high radon levels just below the ground. If your home is on such a spot, and cracks in the basement are letting it in, add at least 250 mrem annually. This is a biggie—your single greatest radiation source. Add 40 mrem if you drink water and eat food. Obviously, this is unavoidable. The most radioactive foods are bananas, potatoes, beer, low-sodium salt (salt substitutes), red meat, lima beans, and Brazil nuts, though the radium in Brazil nuts isn't absorbed by the body.

Add 50 mrem for the natural radiation emanating from within your own body, just as it emanates from all those banana splits and their glowing potassium. (A banana's potassium-40 doesn't really glow. Sorry.)

Add 1 mrem from radiation in the air left over from those atomic tests in the 1950s, conducted in northern Russia, New Mexico, and on several South Pacific islands. If you were alive back then, you knew the world's politicians were screwing around with everyone's health. Just be thankful for the 1963 Limited Nuclear Test Ban Treaty. Before it was signed, our atmosphere's global carbon-14 had doubled, to around fifty tons. Since then, it's returned to almost its natural quantity, which is half that amount.

One side effect of the change in atmospheric carbon-14 is that it has enabled researchers to use a technique called bomb pulse dating for determining the birth year of any individual. That's right—they can tell how old you are simply by measuring the amount of carbon-14 in your tooth enamel and in the lenses of your eyes. Researchers have been using such radiometric dating for the last half century. This method relies on the fact that half of any sample of carbon-14 changes into nitrogen in 5,730 years, and that every living or once-living organism, including the cotton clothing worn by entombed pharaohs, has carbon-14.

Out of every trillion ordinary, non-radioactive carbon atoms in the air and therefore in our bodies, there is one of carbon-14 (^{14}C), which acts chemically like ordinary carbon but has two extra neutrons in its nucleus. When we die, we stop taking new carbon in. The most stable, commonest carbon, carbon-12, lasts forever, but half our carbon-14 is gone in 5,730 years. Half the remaining sample (meaning three-quarters of the original) is gone after two half-lives, or 11,460 years.

So by measuring the ratio of ^{14}C to ^{12}C, we can tell how long ago a plant or animal died.

The atmosphere's ratio of normal ^{12}C to radioactive ^{14}C remains quite steady over time. But thanks to nuclear testing in the late 1940s through the 1950s, far more carbon-14 was suddenly in the air. Back then, the media generally sounded louder concerns about strontium-90 than they did about ^{14}C. And for good reason: fallout from seventeen years of intense atmospheric nuclear testing dispersed strontium-90 throughout the entire globe. However, with a half-life of 28.8 years, around 75 percent of it decayed away by 2017.

Similarly, nuclear bomb testing also released large amounts of cesium-137—which emits prodigious amounts of gamma radiation, the worst kind—into the atmosphere. But with a half-life of thirty years, that, too, has mostly decayed.

We would have had a bigger problem with carbon-14, given its worrisome 5,730-year half-life. But a wonderful mechanism came to our rescue: removal from the air by natural processes. This carbon has mostly been absorbed into the earth and the seas, so we're no longer breathing the bulk of it as airborne radioactive carbon dioxide.

The testing of nuclear weapons between 1945 and 1963 unleashed more gamma rays than the earth had cumulatively received since before the Roman Empire. It also blew fifty tons of radioactive carbon-14 into the atmosphere—roughly double what's present in the air naturally. *(U.S. Army Photographic Signal Corps)*

We also receive tiny amounts of radiation that only true hypochondriacs need to think about. Still, if you fall into that category, be aware that:

You get 1 mrem for each one thousand miles you travel by jet. A single coast-to-coast round-trip gives you 6 mrem. Think of it as frequent-flyer radiation.

Add 40 mrem for each medical or dental X-ray you have. Not much. Again, it's those CT scans that present a significant hazard, especially whole-body scans.

The following sources are examples of radiation so minor that you can safely ignore it when you see it mentioned in some media story designed to give readers or viewers a scare:

Wearing an LCD watch delivers .06 mrem annually.
Living within fifty miles of a coal-fired power plant gives you .03 mrem. (That's because coal and soot are slightly radioactive.)
Having two smoke detectors in the house: .02 mrem.
Living within fifty miles of a nuclear power plant: .009 mrem.
Being in the vicinity of the machine while your airport luggage is X-rayed: .002 mrem.

Scientific debate continues concerning the effect of super-small amounts of radiation. The Mayo Clinic, the National Cancer Institute, the Health Physics Society, and the vast majority of the world's epidemiologists have concluded that very low doses of radiation produce no health consequences at all. None. Zilch. Zero. But other scientists believe that very low doses (in the 1 mrem range) might create some small effect along the lines of one cancer death per forty million people.

So let's say you live in Denver. You have radon in your basement, fly twenty thousand miles a year, and get one full-body CT scan per year. Should you worry? Well, you've probably upped your chance of getting cancer someday by a factor of one in a thousand. But because everyone already has around a 25 or

20 percent chance of getting cancer, the increased risk is still relatively minor.

If radiation concerns you, get your basement tested for radon and, if necessary, install a venting fan. That's a relatively painless way to reduce your radiation exposure by hundreds of millirems. Avoiding unnecessary CT scans, and maybe even skipping a few of those commercial flights, could cut out another 100–1,000 mrems per year.

Oh, yes, one more thing. Don't even *think* about moving to Mars. Martian colonists may receive enough radiation in two years to destroy 13 percent of their brains.

Most of us would probably prefer to use our lifetime radiation allowance a little at a time. Soaking up the tropical sun, say, or flying to Bali. But the time may come when a few of us may squander it in a single shot—for example, by becoming an astronaut on an interplanetary odyssey. A two-year mission to Mars would expose you to more than the government's lifetime radiation allowance for nuclear power plant workers. It's a problem that will have to be solved if ever we are to colonize the Red Planet.

This presents no small problem for future astronauts. Probably the most hazardous environment is Jupiter, which has an enormous radiation-trapping magnetosphere. Its attractive moon Europa, replete with warm salt-water oceans, is the likeliest place for us to look for extraterrestrial life. But on Europa's ice-covered surface, a person in a space suit would get a lethal radiation dose every ten seconds, which is the same as what you get from standing thirty feet from the core of a one-gigawatt nuclear reactor.

But we'll let future generations worry about that.

CHAPTER 16

The Atomic Quartet

It may be argued that there were more fundamental discoveries made during the 1890s than there were during any other decade in history. In 1897, in the excitement following Wilhelm Röntgen's announcement of his discovery of X-rays, the world quickly learned of Joseph John Thomson's discovery of the electron, the first-ever subatomic particle. At this same time, Lord Kelvin and the Scotsman William Ramsay were discovering new elements almost monthly. One of those elements, helium, when heated in a lab and viewed through their spectroscope, displayed a set of bright colored lines that, like a fingerprint, perfectly matched the only unidentified solar emissions up to that point and thus revealed the last unknown substance existing in the sun. Because of this, the new element was named for the Greek god of the sun, Helios. They found that it is also nothing less than the second-most-prevalent element in the universe. Another of these "noble gases" (so called because they don't deign to sully themselves by combining with oxygen or anything else but generally remain in their pure state) is argon, discovered by Ramsay in 1894. Argon, a major component of the air we breathe, makes up nearly 1 percent of the atmosphere and is surpassed in abundance only by nitrogen and oxygen, both of which had been discovered more than two hundred years earlier.

Ramsay even passed high-voltage currents through these gases to create the first neon tubes (neon was yet another of his discoveries), and thus he became single-handedly responsible for the nocturnal surrealism that soon dominated the commercial districts of the world's cities. Ramsay discovered more elements than any human being either before or since.

In the midst of this commotion, several brilliant physicists could scarcely sleep: the recently solved mysteries actually seemed minuscule next to the ones that remained. But many of these mysteries were hovering on the cusp of elucidation. Two of these physicists were a married couple who in a few short years would be known around the world — Pierre and Marie Curie. Another was New Zealand–born Ernest Rutherford, who ultimately revealed the nature of the atom. The fourth member of our 1890s science quartet is Antoine Henri Becquerel, born in Paris in 1852 to a distinguished family of scholars and scientists. Each opened the first doors to our understanding of nature's unseen entities.

Becquerel was a French professor of applied physics; his passion in the 1890s was phosphorescence — the odd emission of colored light after a substance is exposed to light of another color. (Imagine shining red light on a particular odd-looking rock only to see it emit green light after nightfall.) When Röntgen discovered X-rays, Becquerel assumed that phosphorescent materials such as uranium salts might glow simply by emitting some X-ray-like radiation after being stimulated by bright light.

In February of 1896, that watershed year for invisible light,

Becquerel exposed uranium to sunlight and placed it on photographic plates he had first wrapped in thick black paper. Sure enough, when he developed the film, it revealed an image of the uranium crystals. He concluded, "The phosphorescent substance emits radiation which penetrates paper opaque to light." He assumed that the sun's energy, absorbed by the uranium, was then released as X-rays or something similar to them.

Becquerel planned to repeat the experiment on February 26 and 27, but on those days Paris was overcast. With no sunshine to which he could expose his uranium, he put the wrapped-up photography plates in a drawer, along with his uranium. Two days later, on a hunch, he developed the plates, expecting to see only faint images because the uranium had not been stimulated by anything other than dim house light. Instead the images of the uranium crystals were vivid.

There was only one explanation: uranium was independently emitting its own invisible rays. It didn't require any prestimulation by an external energy source. Meaning that some materials spontaneously emit energy on their own. They don't need to be first excited by light or heat. Becquerel initially assumed that this emission was X-rays or some similar unknown invisible light. But when he performed a critical magnetic test to settle the matter, he found that the emission's path was bent by a magnetic field. Neither X-rays nor any other light deflects when passing near a magnet, which meant that the uranium couldn't be emitting any sort of unseen ray. Instead it must be giving off hordes of tiny charged particles, like a swarm of bullets. Becquerel had discovered radioactivity, though he failed to give it a name. For this he was awarded the 1903 Nobel Prize in Physics, which he shared with the Curies.

Hearing of Becquerel's experience with uranium, Ernest Rutherford started to explore its radioactivity and soon changed our knowledge of nature at its most fundamental level. Born in 1871, Rutherford was raised on the mostly rural South Island of New Zealand, the fourth of twelve children. His father, a flaxseed farmer, struggled financially, while his mother, Martha, was a poorly paid schoolteacher. As a child, Rutherford was painfully aware of his family's financial struggles. They even supplemented their income by collecting birds' nests so the family could use the eggs.

Determined to succeed in the world, Rutherford earned a degree at what was then known as the University of New Zealand and was awarded a scholarship to Cambridge. There he became Joseph John Thomson's first graduate student (before Thomson discovered the electron). And it was there that he turned to exploring radioactivity.

Using a more advanced measuring technique than the mere fogging of photographic film—namely, the degree to which radioactivity ionizes the surrounding air so that electric current can penetrate it more easily—he found that uranium as well as thorium emissions have a double component. One component, which he soon named alpha radiation, was absorbed and blocked during his experiments by just a few thousandths of an inch of metal foil, beyond which it was undetectable. But the other component, which he named beta, easily passed through one hundred times as much foil before it vanished.

Moreover, when he subjected each component to magnetic fields, he found that while both their paths were deflected, the beta rays shifted dramatically while the alpha rays barely budged at all. This told him that the beta rays must actually be lightweight particles with an electric charge whereas the alpha particles must be heavy and neutral, or chargeless.

Rutherford later accepted a teaching post at McGill University, in Montreal, where he continued his research. His terms *alpha rays* and *beta rays* became globally accepted in 1899 as a way to describe the two distinct types of radiation.

In 1902, working with uranium, thorium, and radium—a brand-new substance discovered by Marie and Pierre Curie that emitted far more intense radiation than either uranium or thorium—Rutherford and his assistant, Frederick Soddy, formed a theory of atomic disintegration to account for all the oddities their experiments were revealing. Until that year, atoms were assumed to be the stable, eternal, unalterable basis of all matter. But Rutherford and Soddy showed that radioactivity is actually the disintegration of atoms into other types of atoms, along with their components. By definition, in other words, radioactivity is a process of disintegration. Here was a phenomenon the alchemists had tried to reproduce for centuries: one element being changed to another.

In 1903, Rutherford studied a powerful new type of radiation emitted by radium and found that it had amazing penetrating power. This radiation, which had been discovered by the French chemist Paul Villard, was very different from his alpha and beta rays. Rutherford naturally named this third type of radiation the gamma ray. He found it could easily penetrate metal foil; it even survived after passing through several inches of lead. Gamma rays did not deflect—not in a magnetic field, not in an electric field. They always went straight ahead, which meant they must be a form of invisible light rather than particles.

Rutherford did more than merely find and identify various kinds of radiation. Along with the Curies, he described their natures. When the dust cleared, radiation was finally explained. It turns out that heavy radioactive elements have such massive nuclei that pieces of them spontaneously break off. Thus an alpha ray was initially (and correctly) called an alpha particle—a massive chunk consisting of two protons and two neutrons, which commonly exists in nature as the nucleus of helium, the second-most-common element in the cosmos.

Beta particles turned out to be electrons. Simple lightweight electrons, nothing more. They have a negative charge of one.

Rutherford found that a gamma particle is the only type of radiation that is in fact not a particle but a ray of light, with no charge and no weight. History, it seems, saved the strongest for last: gamma rays are not only the final variety of invisible light to be discovered but also the most energetic, with the power to damage any atom or molecule. We'll explore these mighty rays in the next chapter.

In 1909, Rutherford and two assistants, including Hans Geiger of Geiger counter fame, performed the now-famous gold foil experiment. Because gold is so ductile, it can be pressed or hammered into a foil thinner than that of any other metal. Rutherford fired a beam of alpha particles at a layer of gold leaf only a few atoms thick.

At the time, every atom was thought to be like a lump of pudding, a hypothesis proposed by Joseph John Thomson. The atom's negative charge was thought to be scattered like little raisins throughout a positive sphere. But if this pudding model were correct, the atom's positive components would be spread out rather than concentrated, pointlike, in the center. Moreover, alpha particles hurled at foil made up of heavy gold atoms would only be deflected by small angles as they passed through, because the heavy alpha particle would never encounter a massive enough obstruction to seriously alter its high-speed trajectory.

But that's not what happened. The amazing result was that one out of every eight thousand hurtling alpha particles was deflected by a very large angle of more than ninety degrees, while the rest passed straight through with little or no deflection. From this Rutherford concluded that the majority of an atom's mass must be concentrated in a tiny, positively charged region, with electrons surrounding it like planets orbiting the sun.

Many years later, reflecting on his experiment, Rutherford said: "It was quite the most incredible event that has ever happened to me in my life. It was almost as incredible as if you fired a fifteen-inch shell at a piece of tissue paper and it came back and hit you."

Using mathematical analysis, Rutherford proposed, in 1911, a model for the atom that is still accepted today. He concluded that all the positive charge and essentially all the mass of the atom is concentrated in an infinitesimally small fraction of its total volume. This core is ten thousand times smaller than the atom itself. He called this the nucleus, from the Latin for "little nut." The electrons orbit far away from the nucleus and from each other; they're in a mostly vacant realm. Thus the vast majority of the volume of an atom is empty space.

And though this makes us momentarily jump ahead in our story's chronology, it must be mentioned that in 1917 Rutherford proved that multiple copies of a hydrogen nucleus are present in every other atom's nucleus—and he proposed in 1920 that this positively charged component in every atom be called the proton. That same year, he said that an atom's nucleus must also contain a separate, different kind of massive particle, one that has no charge at all—essentially a proton somehow fused together with an electron. He suggested that these theoretical subatomic particles, each with a neutral charge and a mass equal to a proton's plus an electron's, be called neutrons. The name stuck, even if actual neutrons were not discovered for another dozen years.

Rutherford was right about all of it. Today we know that yes, every atom's nucleus consists of protons and neutrons. And yes, this nucleus is tiny, yet it contains virtually all the atom's weight. Each proton is 1,836 times more massive than each electron orbiting it. Yet this heft is confined in an unimaginably minuscule volume.

To be so small yet have such mass means that protons and neutrons are astonishingly dense. To equal their density, you'd have to crush a cruise ship down until it was the size of the point in a ballpoint pen. Imagine a sphere smaller than a mustard seed, weighing what a cruise ship does and containing every ton of its steel. Seems impossible, right? And yet that exact density exists in each of the one hundred billion billion billion protons and neutrons in each of our bodies.

It is true that when Becquerel and Rutherford were making their groundbreaking discoveries about radiation and the atom, the discovery of important underlying mechanisms of physics still lay in the future. Einstein's relativity theories of 1905 and 1915—and the astounding quantum theory of Niels Bohr and Max Planck and its refinements by Paul Dirac, Erwin Schrödinger, and others—had not yet been developed. Nonetheless it is obvious that thanks to these physicists and the Curies, our knowledge of nature made an enormous leap in the years surrounding the turn of the century.

Marie Curie was born in Poland in 1867 as Maria Sklodowska and was forever bound to that country in her heart, insisting that both her daughters become fluent in Polish long after she'd immigrated to France. An autodidact, she spent her adolescence reading books and becoming obsessed with science. In 1889 she started working as a governess in Warsaw, supporting herself while studying at Flying University and finally pursuing scientific training in a chemical laboratory in downtown Warsaw.

She left Poland for Paris at the end of 1891 to live at first with her sister and brother-in-law before finding a tiny apartment in the Latin Quarter and studying chemistry, physics, and mathematics at the University of Paris. Life was difficult; she was almost always destitute and occasionally fainted from hunger.

Studying during the day and tutoring in the evenings, she nonetheless earned her degree in physics in 1893 and was hired by an industrial laboratory. After earning a second degree in 1894, she met the love of her life, Pierre Curie, a young instructor at the Paris Municipal School of Industrial Physics and Chemistry. They bonded over their shared love of science and quickly discovered mutual interests, such as taking long bicycle trips. They were married in the summer of 1895.

Near the end of that year, Wilhelm Röntgen discovered X-rays, and the Curies were swept up in the uproar that followed. Much about this form of invisible light was still mysterious. Discoveries were following one another like boulders in an avalanche. A few months later, early in 1896, Antoine Henri Becquerel showed that uranium emitted rays whose penetrating power closely matched X-rays and that this emission seemed to arise spontaneously from the uranium itself. While the world reeled from these findings, Marie decided to research these new rays herself.

Her first breakthrough came along when she discovered a far more precise way of quantifying uranium's emissions (which Marie, in a published paper, termed *radiation,* a label that permanently stuck) than the simple method of fogging photographic film. More than a decade earlier, Pierre and his brother had developed a sensitive instrument for measuring the strength of an electric charge. Using this device, called an electrometer, she measured the intensity with which uranium rays caused the air around them to conduct electricity. It was a vast improvement over the previous method, and it actually underlies the principle behind the Geiger counter, an invention that would come along later.

Marie's very first results showed that the amount of detected radiation varied with the weight of the uranium, forcing her to conclude that the emissions came from the element's atom rather than from some interaction with external light or other substances. This was the first step in her groundbreaking discovery that atoms are *not* permanent and indivisible but can break down and mutate.

In July of 1898, Marie and Pierre published a paper under joint authorship announcing the existence of an element they named polonium, in honor of Marie's native land. Six months later, on the day after Christmas in 1898, the Curies announced their discovery of a second element, which was an amazing one million times more radioactive than uranium. They named it radium, from the Latin word for "ray."

Alas, radium was present in such minuscule amounts in radioactive ores that it proved nearly impossible to isolate in pure form—normally the first and most essential step in nailing down the properties of any new substance. (The goal is to eliminate impurities that can confound future tests.) Finally, in 1902, using one ton of uranium's main ore, pitchblende, the Curies managed to produce one-tenth of a gram of radium chloride. It then took until 1910 for Marie to isolate pure radium in its metallic state. But it was an important quest for her. She'd made the discovery that when exposed to radium, tumors shrink, and she soon pioneered its medical use, regarding it as a miracle substance. For her whole life, she alluded to it as "my beloved radium."

Pierre died suddenly in Paris on a rainy, windy night in 1906, run over by a horse-drawn carriage whose wheels fractured his skull. Marie, her fame only growing, lived for another quarter century. She succumbed to aplastic anemia in 1934 during a visit to Poland, done in by the very substance—radium—whose name she'd coined. She had routinely carried unshielded vials of radium in her lab coats as they silently emitted the most powerfully penetrating invisible light ever discovered—gamma rays. To the end, she never admitted that radium carried any peril. However, her notebooks, more than a century later, still remain too radioactive to handle. To this day, special suits must be worn to peruse them.

For her work, Marie won the Nobel Prize in 1903, sharing it with Antoine Henri Becquerel and her husband, Pierre. She was the first woman to win the prize and the first person to win a second Nobel, in 1911.

CHAPTER 17

Gamma Rays: The Impossible Light

To harm us, light has to penetrate skin and then damage cells. Visible light can't do this, and neither can radio waves, microwaves, or infrared radiation. But the highest-energy forms of light, X-rays and gamma rays, zoom into our bodies as if our skin were made of mist.

By the final few years of the nineteenth century, scientists were getting better at detecting radiation. At first the fogging of photographic plates wrapped in black paper was the only available way to see these emanations. Using this technique was a quick and easy way to determine which substances did or did not create radiation. For example, no fogging of film ever followed exposure to Herschel's "calorific rays" or Hertz's radio waves. Even Ritter's "chemical rays" (ultraviolet rays), despite their higher frequency and greater energy, didn't leave markings on sealed photography plates. But Röntgen's rays did. They alone—along with the soon-to-be-discovered gamma rays—had sufficient penetrating power.

By the time the Curies discovered radium, in 1898, more refined methods of detecting unseen emissions were available. As they pass through air, charged particles and high-energy light knock electrons from the neutral atoms of gaseous oxygen and nitrogen, leaving them with an electrical charge. This charge, in turn, can be measured by its ability to conduct electricity through the gas.

We've seen that in 1899, Rutherford named Becquerel's easily blocked radioactive emanations alpha particles and labeled those with greater penetrating power beta particles. In 1900, Paul Villard, a forty-year-old Parisian schoolteacher who maintained a small laboratory, started to study the emissions of the element radium. He put a piece of radium in a lead box that had a small opening from which its "rays" would stream. He saw and recognized the previously described radium rays and realized that something new was also streaming out, something more powerful and penetrating than anything previously observed. A modest man, Villard did not give this superpower emission a new label but merely described its properties. Three years later, in 1903, Ernest Rutherford named Villard's discovery gamma rays to keep his Greek-letter nomenclature system intact.

It took years to figure out what they were. Rutherford first assumed they were alpha particles that somehow had a much higher speed and that this was why they could penetrate dense materials such as metal. But while alpha particles were slow to deflect in a magnetic field, gamma rays would not deflect in any way. That was an important clue that they might be a form of light and not a particle at all. Finally, in 1914, Rutherford was able to bounce them off crystal faces and measure their wavelength. Bingo: gamma waves, the size of atoms, were so incredibly tiny and had frequencies so incredibly fast—more than ten million trillion per second—that they had more energy than even X-rays.

Here was the most powerful form of light ever discovered—a distinction gamma rays retain today. The power of gamma rays (also known as gamma radiation) is so great that it rips apart almost everything it touches. It is the most hazardous form of invisible light.

Alone among the various kinds of invisible light, gamma rays do not have a cut-and-dried wavelength. At one time they were categorized as waves with a shorter length and higher frequency than X-rays. Astronomers still use that as a hard-and-fast definition, no matter where the rays come from or how they are created. But some physicists now categorize gamma rays according to their origin alone. Gamma rays are usually born not as a result of electrons changing their positions, the way all other light is created, but from an atom's nucleus as it emits energy when in an excited state (when it's undergoing fission) or when it suddenly changes shape. Such nuclear transition occurs in a trillionth of a second, making gamma-ray production an extremely short-lived affair. A common modern definition holds that if rays emanate from an atom's nucleus, they are gamma rays, period. If they originate from the motion of electrons, then they are some other variety of light. This controversial distinction has allowed some overlap between the frequencies we call gamma rays and those we call X-rays.

The bulk of the universe's gamma rays are created by unimaginably violent events that take place far from our world, such as the annihilation of antimatter (matter with all its electrical charges reversed) when it strikes regular matter. Or supernova explosions. Or collapsing supermassive black holes.

Among the most exotic of these events are the gamma-ray bursts, discovered in the 1990s. Somewhere in the sky, beyond our own Milky Way galaxy, an average of once a day, a long (meaning ten-to-fifty-second) stream of gamma rays emanates from a single spot in the heavens. This is now believed to be the result of distant collapsing supermassive black holes, each emitting more energy per second than the sun gives off in its entire lifetime.*

A bizarre gamma-ray discovery in 2010 continues to cause much consternation. In November of that year, astronomers using NASA's Fermi Gamma-ray Space Telescope, launched in 2008, announced something truly astonishing. Emanating from the center of our Milky Way galaxy are two bubbles made solely of gamma rays.

This would have been strange enough if the bubbles, expanding at 2.2 million miles an hour, were concentric—a bubble within a bubble—and were both centered at our galaxy's core. But no. The two enormous spheres each hover in seemingly empty space above and below the black hole in the Milky Way's nucleus. They are tangential to each other, touching at the galactic center to form a squat hourglass shape. The entire structure looks like the number 8.

---

* A black hole is a place of extreme density where gravity has become so strong that any object would need to move faster than light, which is impossible, in order to escape. Since light cannot escape, the object appears black. In practice, the only objects that can collapse to this degree are unusually massive stars in their old age, when their core hydrogen fuel has been so depleted that it can no longer supply enough outward-pushing energy to counterbalance the in-falling weight of the layers above it. In the heart of every galaxy are supermassive black holes, whose mass is not merely ten or twenty times that of our sun but millions or even billions of times larger. The relevance to our story is that nearby subatomic particles and atoms can be pulled into a super-high-speed orbit around both stellar and supermassive black holes to form a so-called accretion disk. Just before falling in and vanishing from our sight, such particles will be moving at such high speeds that they'll be stimulated to emit high-energy bits of light, such as gamma rays.

An edge-on view of our Milky Way galaxy, showing the enormous, mysterious, ultrapowerful gamma-ray bubbles discovered in 2010. *(NASA)*

Stars do not emit gamma rays. This is why that dense gamma-ray swarm at our own galaxy's center is so puzzling. It's the unmistakable sign of extreme violence. And yet these days the Milky Way's core is about as energetic as a steamy July lunchtime in New Orleans.

The bubbles are sharp-edged, well defined, and enormous. The top and bottom of the figure 8 extends from twenty-five thousand light-years north of the galactic plane to the same distance beneath it. From our earth's sideways viewpoint, twenty-five thousand light-years from the center, the hourglass stands a whopping forty-five degrees above and below the galactic core in Sagittarius. It takes up half our southern sky.

Theorists need to explain more than just what could have produced this kind of extreme energy, which is equivalent to one hundred thousand exploding supernovae. They must also explain the off-center nature of the bubbles, because each seemingly surrounds nothingness.

Imagine a small expanding bubble inside a larger expanding one, both of them centered on the giant black hole at the center of our galaxy. We'd assume that some violent event in the past caused both bubbles to form—especially since both are composed of highly energetic gamma rays. But that's not what we see. Instead we observe two separate bubbles, one atop the other, as if they were drifting and got stuck when they made contact. This place of bubble contact is the center of our galaxy, where the supermassive black hole floats. Yet the centers of both bubbles are high above and far below the galaxy's center, in a region where we detect nothing at all. How strange it is!

Jon Morse, former director of astrophysics at NASA headquarters, summed up the discovery at a press conference: "It shows, once again, that the universe is full of surprises." This gargantuan hourglass, called the Fermi Bubbles in honor of the orbiting gamma-ray telescope that found them, is now regarded as an entirely new type of astronomical object.

In trying to come up with some explanation for our galaxy blowing gamma-ray bubbles at temperatures of seven million degrees Fahrenheit, many astrophysicists express a gratifying unanimity: they say, "We have no idea." Others, starting perforce from square one, have posited a couple of vague possible causes. The first is that, perhaps a few million years ago, a burst of star formation at the galactic center created numerous massive stars, all with high-speed winds consisting of high-energy particles. Since this alone could not begin to explain the superhigh energy within the bubbles, that theory further imagines that many of these stars blew up into supernovae simultaneously.

Don't like that one? Neither do I. So let's go to possible explanation number 2, which is that the four-million-solar-mass black hole at our galaxy's center had a brief feasting frenzy when it captured particles shed by stars that once lurked at our galaxy's core. This captured material was accelerated at its accretion disk, where material remains temporarily visible before falling in to the innermost zone, from which nothing can escape. Then, perhaps, that black hole could have developed something it does not presently have: twin jets of outrushing material. We see such blue jets exploding from the supermassive black holes in a few other galaxies. These jets could have possibly deposited energetic material above and below the galactic plane, although how bubbles then emanated from those positions is anyone's guess.

The answer could be even stranger. Might these be long-sought signs of dark matter, the hypothesized material that possesses a gravitational pull and yet remains utterly invisible? Could this be what's making our galaxy spin as if it were a solid vinyl record and causes other galaxies to behave oddly, too? Could dark matter be meeting its opposite entity (whatever that is) in total annihilation, the way matter and antimatter do? Maybe. More likely, though, is that the double bubble is something else entirely, some new phenomenon that will actually get in the way of the dark-matter hunt.

As Fermi research team leader Douglas Finkbeiner put it in an interview, "This just confuses everything."

Fortunately, few gamma rays reach us here at the earth's

surface. An orbiting gamma-ray telescope, when pointed downward toward the ground, revealed that only five hundred brief flashes of gamma rays appear on earth each day, mostly from intense lightning storms.

Gamma rays are briefly emitted in the explosion of atomic and hydrogen bombs. But in terms of practical use, the fact that they easily destroy life—even insects, which are generally highly radiation resistant—has led the food industry to push for food irradiation as a way of preventing the sprouting of vegetables and fruits, thus preserving their flavor. The process is already widely used to sterilize spices. In practice, this means that rolling shelves containing these products are exposed to high-intensity gamma rays, then rolled away as though they were on an amusement park ride. This kills all insects and pathogens.

Those in the natural food movement are not happy about food irradiation. They maintain that all foods contain a scientifically undetectable "life force" (called *prana* in India) that is destroyed by the process. Irradiated fruit, according to this view, may still contain its original quantifiable vitamins, and it may look the same as other fruit, but its cells and vitality have been destroyed by the gamma rays so that it lacks any "spark of life."

Radium, used as a source of gamma rays, has also been employed to treat eczema and other skin rashes and to remove benign skin tumors and moles. Such radiation treatments were administered for almost forty years starting in the 1920s. They were also used to treat enlarged thyroid glands, inflamed tonsils and adenoids, asthma, whooping cough, and even a mother's lactation problems after birth. These days, the one remaining acceptable medical use for focused gamma-ray beams is the destruction of tumors, mostly as a form of palliation.

* * *

Now you have heard the stories of how various forms of invisible light were discovered. The discoveries spanned nearly a century: the first invisible light was detected by the self-taught astronomer William Herschel in 1800; the most recent was identified by schoolteacher Paul Villard in 1900.

Throughout history, new elements and compounds have been found and their properties analyzed and exploited. But we have now seen that over the course of one curious century, a coterie of singular men—and one woman—made a series of discoveries that were truly bizarre. The sun, along with seemingly ordinary metallic-looking elements, sends out invisible rays, or in some cases tiny invisible bullets. These ghosts race through the air and affect other objects. They can even influence living beings. It was as if the scientists had discovered phantoms among us. As the technology needed to produce these invisible energies and use them in various products and applications became available, we soon reached a situation in which they fill all the spaces on our planet and fly through our bodies nonstop.

How do these phantoms affect our lives and our health as they provide their almost magical conveniences? We need to find out, since one kind of invisible light in particular lies at the center of our lives.

CHAPTER 18

Cell-Phone Radiation

We can't say that no one saw the cell-phone revolution coming. In 1959, nearly a decade before his *2001: A Space Odyssey* came out, science visionary Arthur C. Clarke wrote an essay describing a "personal transceiver, so small and compact that every man carries one." He foresaw a time when "we will be able to call a person anywhere on earth merely by dialing a number." The device would even contain some sort of global positioning system so that "no one need ever again be lost." In a later book, *Profiles of the Future,* he predicted such telephones being available by the mid-1980s. (Yes, he was off by a decade. Clarke usually nailed the specifics but messed up on the timing. He envisioned humans venturing to Jupiter by the year 2005.)

We've come a long way from those huge brick-size satellite phones that were introduced in the 1990s to our present time, when virtually everyone has a personal phone in his or her pocket. Movies from the 1990s still show people waiting in line to use pay phones at the airport. These days, for a mere $150, residents of villages in South India that are still entirely off the grid can purchase siding-mounted exterior solar panels that let them charge their cell phones.

We have microwaves to thank for that progress. They're the

medium that makes cell phones work. And unlike Percy Spencer's microwave ovens, the new devices are not the technological brainchild of any one person. The idea of cordless telephony has been around for nearly a century—some primitive systems were installed in European trains even before World War II. After that came "radiophones," which often took up an entire attaché case and ultimately got as small as a milk carton, though they still didn't allow the user to roam from one area to another.

The concept of a cellular network within which signals could transfer from one tower to another was developed in the 1960s through pioneering work by Bell Labs engineers Richard H. Frenkiel, Joel S. Engel, and Philip T. Porter, particularly the latter, when they suggested that cell towers use the now-familiar directional antennas, which would minimize interference and allow the reuse of specific microwave frequencies. Porter also was the first to suggest the dial-then-send calling method, which was ultimately adopted and which prevents wasted "on-air" channel usage.

In the 1970s, Bell Labs engineer Amos Joel invented a three-sided trunk circuit to aid in the call-handoff process from one cell to another, but switching improvements—meaning methods of sending and receiving calls—during the next decade leaped over even this idea and went directly from circuit switching to packet switching, in which entire chunks of data seamlessly change their transmission points. This is not the place for the almost magical story of technology's evolution from the early 1G cell phones of the 1990s to the 2G, 3G, 4G, and 5G phones of later years, each making great leaps in data-carrying capabilities and bit rates to the point where intensive data-streaming capacity is now routinely expected on one's smartphone.

Instead we are here concerned about the invisible rays that make such streaming possible and what they may do to our bodies in the process.

We must also note that microwave signals originate not just from towers but also from satellites in high orbit, at 11,300 miles. The combined radio-spectrum environment from, among other things, the two dozen American GPS satellites now in space, new Chinese, European, and Russian satellite systems, and Iridium satellites, which provide satellite phone service, ensures that our homes and bodies are never devoid of electromagnetic radio waves. But if we're near a cell tower, or in a Wi-Fi hot spot, we're positively awash in them.

Visualizing microwaves isn't easy. Sizewise, picture each ray as a curled-up caterpillar, with its back raised and its front and back ends lower down. The largest microwaves look more like garter snakes, a foot in length, again with the midsection raised in a curve. Each curved caterpillar vanishes, then a new one appears in the same spot a trillion times a second. That's a million million. It's fourteen thousand times faster than individual frames appear in modern movies.

Should we be concerned? Some people certainly do worry. You can read the anxiety in local newspapers, where letters to the editor express fear about microwaves from cell towers, Wi-Fi, and even cell phones themselves and urge that Wi-Fi be banned in schools. Authorities sometimes respond to this pressure: in Woodstock, New York, in 2015, the school board placed a temporary restriction on school microwave Wi-Fi installations pending further investigation.

The issue of whether cell phone radiation is putting us in danger has generated much press, some thoughtful, some paranoid. An example of the latter is the widespread belief, expressed all over the Web, that corporations involved in the cell-phone industry (including manufacturers, electronics and software companies, and the carriers themselves), as well as government regulatory agencies and even large mainstream health organizations such as the American Cancer Society and the Mayo Clinic, are actively conspiring to suppress microwave hazards of which they are supposedly well aware.

Some of the fears are based on a report issued in 2011 by the World Health Organization's International Agency for Research on Cancer (IARC). The agency had gathered in Lyon, France, to discuss scientific studies surrounding the question of whether there's a relationship between radio-frequency-modulated electromagnetic fields (RF-EMF) and cancer. After intense deliberations, and to the great surprise of the world at large, experts decided to classify RF-EMF waves emitted by cell phones, cell towers, and Wi-Fi networks as category 2B, indicating a "possible human carcinogen."

On the other hand, as a *New York Times* article pointed out in 2016, there have been many studies conducted on the issue, including the Million Women Study, in Britain, a Danish study of more than 350,000 cell-phone users, and studies examining the effects of radio waves in animals and cells growing in petri dishes. Those studies indicated that there is still "no convincing evidence of any link between cellphone use and cancer or any other disease. Also, the incidence of brain cancer in the United States has remained steady since 1992, despite the stark increase in cellphone use."

Between the alarming conspiracy theories, the "possible human carcinogen" verdict, and reassuring reports like the one in the *Times,* it's hard to know what to believe, and that in itself can cause anxiety, because we certainly need to know. More than a billion people use cell phones daily. Are our phones putting us in danger or not?

The short answer is *probably* not, but it's still better to take certain precautions.

First of all, the microwave and radio bands consist entirely of nonionizing radiations. They simply don't have the energy to knock electrons out of their orbits, which means they can't cause changes on an atomic level. So even with prolonged exposure to microwaves or radio waves, there's no danger of gene mutations or chromosome damage. And presumably there's no possibility of such rays being carcinogenic.

Indeed, microwaves lie not merely outside the ionizing part of the electromagnetic spectrum but also far from it. Microwaves are less energetic than infrared radiation, which in turn is less energetic than visible light. And nobody can be harmed by, say, red mood lighting, even if it's bright.

Is it okay to warm up your teenager's brain? Regardless of whether microwaves can cause cancer (we'll get to that in a second), we know that they do make atoms jiggle faster—another way of saying they heat tissue. One study showed a measurably increased blood flow on the side of the head where a cell phone was held. The effect was undeniable. But was it deleterious? Might it simply mean that a certain part of the brain had become actively engaged? One might point out that drinking a bowl of soup or enjoying a cup of tea will heat far more tissue, and to a much greater degree, than talking on a cell phone will, yet we don't worry about the potential health risks of frequent tea drinking. Moreover, taking a single hot shower heats far more body tissue in one shot than using your cell phone non-stop for a month. So this "heating tissue" business is a good example of a real effect caused by radio-frequency radiation that sounds scary but in all probability is inconsequential.

Now to the cancer question: as of 2016, there are more than seven thousand major studies of RF "radiation" in the medical literature. As the *Times* reported, the very largest studies have failed to detect an association between cell-phone use and brain tumors or other cancers.

The largest investigation is the Interphone Study, which involved thirteen countries, including Canada, the United Kingdom, Denmark, and Japan. Researchers questioned more than seven thousand people who had been diagnosed with a brain tumor as well as a control group of fourteen thousand healthy people about their previous cell-phone use. The study found no association between cell-phone use and glioma (cancerous brain tumor) rates except in the group of participants who reported using their cell phone for at least 1,640 hours in their lifetimes without a headset. Those participants were 40 percent more likely than those who never used a cell phone to have a glioma. Since this finding contradicted other studies that uncovered no increased cancer risk, the Interphone Study authors speculated that people with brain tumors, looking for an explanation for the tragic disease that had befallen them, might be more likely than healthy people to exaggerate their cell-phone use.

Also reassuring are the results of studies involving workers whose occupations expose them to more than a thousand times more RF energy than the rest of us get. These lab technicians, cell-phone-tower maintenance workers, radar technicians, and others show no increased cancer rate whatsoever.

Still, some research leaves the door open to doubt. The results of ongoing studies, in progress since 2013, that expose animals to various microwave intensities have so far been generally reassuring, but in 2016, a study conducted on rats exposed to high levels of cell-phone-type rays for more than two years, starting before birth, found a 2 percent rate of brain cancer—but only in males, not in females. Oddly, none of the control-group rats developed tumors, though the usual rate would have been 2 percent. In other words, if it weren't for an abnormally low cancer incidence in the *control group,* the cancer rate for the exposed rats and the unexposed rats would have been the same, and microwaves would have been given a clean bill of health. The whole thing was puzzling—so puzzling that most researchers do not accept the results, although some do.

One Danish study found that brain-tumor incidence increased among the segment of the population that used cell phones the most hours per day. It didn't help when the giant British insurance company Lloyd's, in 2014, announced that it would no longer sell insurance against health effects from microwaves. Many started wondering whether they were damaging themselves and their children by permitting unrestricted cell-phone usage.

These puzzling outlier studies need to be acknowledged, and we need to continue research into cell-phone radiation. But the fact remains that there has been no convincing evidence to date that cell-phone use increases the risk of cancer. So why the IARC 2B classification of microwaves as a "possible carcinogen"? Well, context is important. After all, the WHO classifies coffee in the 2B category, despite some investigative organizations such as Consumers Union saying that coffee is actually healthful. If after twenty years and seven thousand studies researchers had instead found evidence that microwaves are "probably" carcinogenic, that would have earned them a 2A classification, still lower than a class 1 "definite" cancer-causing rating. In other words, the 2B designation indicates that any effect must be *very* subtle. Indeed, the way things look now, the worst we might eventually find out about microwaves in terms of carcinogenesis is a tiny effect along the lines of one case per several million users—a hazard that would probably not inspire anyone to change his or her habits. Meanwhile, cell-phone-signal emissions have been steadily decreasing since 2005 as the technology has improved, rendering obsolete any findings from studies conducted before that.

As the American Cancer Society wisely points out, the fact that most studies so far have not found a link between cell-phone use and the development of tumors is unlikely to end the controversy and put us completely at ease—nor should it. These studies suffer from a number of limitations, which the ACS lays out: "First, studies have not yet been able to follow people for very long periods of time. When tumors form after a known cancer-causing exposure, it often takes decades for them to develop. Because cell phones have been in widespread use for only about twenty years in most countries, it is not possible to rule out future health effects that have not yet appeared. Second, cell phone usage is constantly changing. People are using their cell phones much more than they were even ten years ago, and the phones themselves are very different from what was used in the past. This makes it hard to know if the results of studies looking at cell phone use in years past would still apply today."

Frequency, intensity, and duration of exposure can affect the response to radio-frequency radiation (RFR), and these factors can interact with one another and produce various effects. In addition, in order to understand the biological consequence of RFR exposure, one must know whether the effect is cumulative or whether compensatory responses result. In short, the issue of whether there is any adverse biological effect from the entire radio-frequency band (which includes TV and radio towers and not merely cell-phone microwaves) is complex. Major study results will be announced between 2017 and 2020, so the last word about microwave safety is still to come as of this writing.

While we wait for that last word, why not do what we can to minimize our exposure? Like all electromagnetic radiation, both visible and invisible, RFR intensity falls off inversely with the square of distance. This means that if you step twice as far away from a lightbulb as you were in the first place it will appear 2^2, or four, times dimmer. Or if you spend your days wondering how bright the sun appears from Saturn, which is around ten times farther away from it than Earth is, simply calculate the square of ten. Thus from the ringed planet, the sun appears one hundred times dimmer than it does from Brooklyn. Quick and easy.

Similarly, if instead of holding your cell phone tightly against the side of your head, a mere inch from your brain, you put it on speakerphone or use a headset so that the phone and its antenna are now in a pocket twelve inches from your brain, you have reduced your brain's incoming microwave intensity by 12^{12}, or a factor of 144. Or you could just join the under-twenty-five generation and switch to texting rather than talking. That way, you eliminate any hazard.

CHAPTER 19

Cosmic Rays

When the thirty Apollo astronauts—in groups of three, with three "repeats"—sped outward from Earth between 1968 and 1972, they experienced something no human had ever been subjected to, before or since. They ventured beyond our planet's magnetosphere—its protective magnetic field.

The results were unexpected and bizarre. Each man saw something that resembled a streaking meteor cross his field of vision around once a minute. At first the astronauts kept this disquieting development to themselves. Nearly all of them were navy pilots, and long experience made it an unwritten rule that no pilot ever reveals to any physician that anything is medically wrong with him. Especially something that might be construed as mental in origin.

But a few were close enough to their fellow astronauts to confide in them about what was happening. In this manner they came to realize that the streaking-meteor phenomenon was befalling them all. Then it was safe to report it to mission control.

NASA physicians had an immediate theory that was later verified. Powerful cosmic rays were zooming through the astronauts' eyeballs. Traveling beyond both Earth's atmosphere and its magnetosphere meant that those high-speed intruders from deep space had nothing to block them. Each "ray" was ripping a path through the cerebral cortex, exciting and no doubt damaging neurons and triggering the streaks.

It was a twenty-nine-year-old Austrian physicist named Victor Hess who discovered cosmic rays. Born in Austria in June of 1883, Hess earned his PhD from the University of Graz in 1906. He first decided to study optics under famed physicist Paul Drude, the man who gave us the symbol c for the speed of light. Tragically and inexplicably, Drude committed suicide a few weeks before Hess was due to arrive.

Hess instead accepted a teaching position at the University of Vienna. The discovery of radium by the Curies in 1898 had created a global sensation, and Hess began a serious study of that hottest issue in physics. Working as an assistant at the Institute for Radium Research at the Austrian Academy of Sciences, he became fascinated by a curious phenomenon: electrical charges were regularly detected inside electroscopes even when no radioactive elements were nearby, no matter how well those containers were insulated. The accepted explanation at the time was that earthly minerals such as quartz and granite emitted periodic radiation that caused such readings. If this were so, then the number of charges inside the device should diminish as one raised the electroscope farther off the ground.

There were good hard-core physics reasons why this should be so. The intensity of light or any other electromagnetic radiation, as we saw in chapter 18, is inversely proportional to the square of the distance from the source. So if a radioactive bit of radium is twice as far away as it had been when it was last observed, you'd receive only one-quarter of its energy in the same time period. A widely accepted scientific paper spelled it out: assuming an even distribution of radioactive rocks on the earth's surface, at an elevation of ten meters, or around three stories, the measured radiation should fall to 83 percent of its value on the ground. A height of ten stories should reduce it to 36 percent, and at a height of one thousand meters, or around three thousand feet, only 0.1 percent of the initial value should remain.

So what accounted for the charges that Hess observed inside the electroscope? The answer lay in research that was just coming to light at the time. A few scientists were finding that radiation did not necessarily diminish with distance from the ground. For example, in 1910, Theodor Wulf took electroscope readings at both the bottom and top of the Eiffel Tower and found that there was far more ionization—i.e., radiation—at three hundred meters (the top) than one would expect if this effect were solely attributable to ground radiation.

Could a major source of the ionization in Hess's electroscopes be the sky rather than the ground? Hess first calculated that at a height of just 1,500 feet, enough insulating air should intervene to prevent *any* ground-based radiation from being detected. Then he mounted his instruments in a balloon, climbed in after them, and took a series of ionization measurements in ten ascents over three years, starting in 1911. He got the same results each time. Radiation activity first diminished as his balloon ascended but then started to rapidly rise. At an altitude of seventeen thousand feet, or just over three miles, the readings were always at least twice as great as they were at the surface. In a published scientific paper, Hess announced that "a radiation of very high penetrating power enters our atmosphere from above."

Hess was no coward. He conducted a perilous flight at night

to eliminate the sun as a cause of the radiation. Sure enough, his readings remained just as strong after nightfall. He also went up on April 17, 1912, during a near-total solar eclipse, when most of the sun's energy was blocked by the moon. Again the radiation intensity did not decrease.

If the radiation Hess was picking up wasn't coming from the sun or from earthly rocks, it must be coming from deep space. At a major 1913 science convention, the outer-space origin of these rays was generally accepted, but they were believed to be gamma rays. More than a decade later, Hess's findings were confirmed by Robert Millikan, who dubbed the mysterious radiation "cosmic rays." In 1936, Hess won the Nobel Prize in Physics, an acknowledgment of his discovery.*

Turns out that Millikan was wrong to assume that these high-energy rays were a form of invisible light. In 1927, investigators began finding evidence that cosmic-ray intensity varied with one's distance from the equator. This wouldn't make sense if the rays were a form of light. But it would be reasonable if they were being deflected by our planet's magnetic field. Thus they must be some kind of charged particle and not any kind of photon or ray. Nonetheless the word *ray* tenaciously stuck and remains in common use today.

Then in 1930, an even weirder phenomenon emerged. Scientists started seeing a difference in intensity between cosmic rays

* Hess, whose wife was Jewish, received threats from the Nazis, so he immigrated to the United States, where he became a professor at Fordham University. He continued testing for radiation after Hiroshima—conducting measurements from the eighty-seventh floor of the Empire State Building, in New York City. He also measured the radioactivity of granite in New York's 190th Street subway station and continued to contribute to the field of physics until his death, in 1964.

arriving in our atmosphere from an easterly direction and those streaming in from the west. This "east-west effect" indicated that not only do cosmic-ray particles have a charge, the charge must also be positive, since the direction through the earth's magnetic field would have the opposite effect on a negatively charged particle. It all meant that cosmic rays are mostly protons, or hydrogen nuclei, whose charges are positive.

By the end of World War II, researchers had more or less determined the true composition of cosmic rays. It wasn't pretty. Ninety percent of them are indeed simply protons—the nucleus of every hydrogen atom. Slightly less than 9 percent are alpha particles, which, as we'll recall, are helium nuclei, meaning a hefty glob containing two protons and two neutrons. One percent are ordinary electrons, or beta particles. And a ragtag assortment of other stuff, including antimatter, makes up nearly another percent.

We'll get to antimatter later. For now, suffice it to say that the cosmic-ray ingredient list made no sense back then and remains just as bewildering today. True, protons and electrons are both expelled by the sun. But the superhigh energies of many cosmic rays rule out the sun as a likely source. So right from the beginning, astrophysicists assumed correctly that most come from supernova explosions far, far away. More recently, other violent celestial events, such as the explosions at the cores of some galaxies and the collapse of black holes, have also been pinned down as cosmic-ray sources. But supernovas and their remnants, such as the famous Crab Nebula, in Taurus, are the main cosmic-ray sources.

Why their proton-heavy composition? There are just as many electrons in the universe as protons. Shouldn't electrons be hurled outward by supernovas, too? What happened to them? Worse, a small percentage of cosmic rays are ultrafast, with unbelievable energies. Some of these subatomic bullets can deliver the same wallop as a baseball hitting your head at fifty-six miles per hour. Imagine: a single particle far smaller than an atom smashing anything it encounters with palpable impact. What would *that* do to one of your DNA strands? What about the particles' overall effect on human health?

If we broke down the radiation each of us is exposed to every year, we'd find that half of it does come from the ground—at least for those whose homes have radon leaking in from cracks in the basement. A small percentage emanates from within our own bodies, from radioactive carbon-14 and the potassium-40 in foods such as bananas. But around one-tenth of the radiation we receive penetrates us from above—and consists of these cosmic rays. As Hess discovered, they grow more intense the higher up we go. So those who live in high-altitude cities such as Denver get fully one-quarter of their annual radiation from cosmic rays alone.

It's even worse for those who spend their days higher than Mount Everest—career pilots and flight-crew members. Thanks to cosmic rays, they get twice as much radiation as people in other professions do, which translates into a 1 percent higher cancer rate than the general population. For the rest of us, to exceed the radiation limit established by the US government for nuclear power plant workers, you'd have to fly more than eighty-five thousand miles a year.

Our exposure to cosmic rays depends not only on location but also on timing. Much of the cosmic radiation streaming in from deep space is ordinarily deflected by a boundary at the edge of the solar system, a shock wave where the outrushing solar wind goes from supersonic to subsonic. But the sun's power varies with its eleven-year solar cycle, during which its storms and subatomic emissions, called the solar wind, alternately grow more intense and then less so. In years when the sun is wimpy, this termination shock zone becomes weaker and its protective barrier much more porous. That's when deep-space cosmic rays stream to Earth with far greater intensity. At such times, and during particularly powerful solar flares, when the sun itself cranks up its cosmic-ray intensity, jetliners flying polar routes are bombarded with extra radiation.

During those times, if you have a camcorder and look at its black screen while you're in a dark closet, you'll see flashes as cosmic rays strike the camera's CCD chip. At such times you can feel lucky you're not an astronaut—or a future Martian colonist—in a flimsy spacecraft outside Earth's atmosphere and magnetosphere. You'd be particularly vulnerable during the half-year-long trip to Mars, when a severe cosmic-ray bombardment could produce truly hazardous radiation levels. You'd see more than mere streaks across your visual field. You might then be silently condemned to a short life span.

By the late 1940s and early 1950s, cosmic rays were deemed a serious hazard to any living organism that ventured beyond our atmosphere. What did that mean for human space flight? To find out, the US Air Force, in a project led by Captain David Simons working under Colonel James Henry, first sent organisms such as fruit flies and mice, and then primates, to the upper regions of the atmosphere in captured German V-2 rockets.

The first monkey flight was set for June 18, 1948, but the animal suffocated in the capsule before it had even left the ground. A year later, another monkey was sent aloft in a better-ventilated capsule. Sadly, this time the parachute failed, and so the creature was just as dead as its predecessor.

By 1954, researchers had switched to high-altitude balloon testing, but the hazard still remained unknown. The animal of choice was often an odd kind of black mouse, because the pigmentation in its hair follicles would be destroyed by radiation, which would immediately turn them white. But good news: these mice stayed predominantly dark. By the following year, Simons had completed dozens of test flights, and the animals, mostly Java macaques, were returning to Earth unharmed. It was therefore only in 1955 that cosmic rays were finally discovered to be nonlethal for future astronauts—at least during missions of relatively short duration. In 1957, just a few months before the Soviets shocked America with the first-ever launch of their orbiting satellite *Sputnik 1,* the air force sent test pilot David Simons up in a balloon to an altitude of over one hundred thousand feet for more than twenty-four hours—along with radiation monitors—without any ill effects. Thus it was only as the curtain was actually rising on space travel that we finally got the last-minute green light for short-duration human missions.

But just because cosmic rays didn't cause rats, monkeys, or astronauts to drop dead on the spot doesn't mean they can't mess with your health. Cosmic rays do something curious as they fly into our atmosphere. Striking air molecules at a height of around thirty-five miles, they break atoms apart the way a cue stick breaks up a rack of billiards. The atoms' contents rain down toward the surface at nearly the speed of light. Among the detritus flying off are muons, which are sometimes regarded as cosmic rays themselves. These strange transient particles are neither superheavy nor particularly lightweight. Rather, they each have a mass equal to around 208 electrons. They also have a short life span. With a half-life of just two-millionths of a second, a muon quickly disintegrates and vanishes, leaving behind a swarm of ordinary electrons and some neutrinos that essentially weigh nothing. Harmless stuff.

But before they vanish, muons can indeed cause harm if they strike the wrong bit of genetic material in a cell nucleus. This is indeed one of the causes of the "spontaneous" tumors that have always plagued the human race. Nor can you hide from them. Some 240 muons flash through your body every second. More hit you if you live in hazardously high-up Denver, and none reaches you if you've chosen a home deep underground or spend most of your time in a subterranean parking garage.

Muons reveal something truly strange. You see, if you do the math, you'll notice that because they are created thirty-five miles up, zoom downward at almost the speed of light, and live for just two-millionths of second before vanishing, they should not logically penetrate our bodies at all. Even at an ultrahigh velocity, no object can travel more than a few city blocks in two-millionths of a second. So how can muons possibly make it all the way here to the surface of our planet?

This is the sort of thing that was utterly inexplicable until Einstein created his two relativity theories, in 1905 and 1915. Only then did we learn that the passage of time and the distance between objects warp and mutate according to local conditions. Thus muons at their high speed would observe (if they were conscious) that the distance between them and the ground is not thirty-five miles but rather just a single city block. The gap between the upper atmosphere and the ground has dramatically shrunk, enabling our muons to arrive at the surface of the earth before their two-microsecond life span has elapsed.

But we observers see something very different. To us, the

distance between the place where the muons were first created and the spot where they strike the ground is not a mere city block but rather thirty-five miles. We observe no change in the muon's path or the distance it traverses to arrive in our bedrooms. However, our measurements uncover something else: we see the muons experiencing a slowed-down rate of time. Their lives unfold in slow motion. And because time is passing at a glacial rate for the muons, their rate of decay is similarly retarded so that they no longer vanish in a couple of microseconds. Rather, their half-life has been stretched out, and they now live so long that they are able to reach the ground during their newly enhanced lifetimes.

We see their time slowed down because their decay happens far more slowly. Muons, by contrast, feel their time passing normally but experience the distance in front of them as contracted. It's exactly what Einstein predicted. Neither space nor time is absolute: they both warp and mutate depending on local conditions, such as speed, which is why he named his theory relativity.

To use real numbers, at 99.9999999 percent of light's speed, a muon experiences the distance in front of it contracted by a factor of 22,361. The original thirty-five-mile gap between it and the earth's surface has suddenly shrunk to a mere eight feet. And our muon says, "Heck, I can cover eight feet within my short lifetime." And it does.

We, on the other hand, observe the muon's time contract so much that each hour shrinks to just one-sixth of a second. From either perspective, the muon gets here and penetrates our bodies. It's just that we and the muon disagree about the amount of distance it covered and how much time it took.

That was Einstein's whole point: neither of us is right or

wrong. There simply is no absolute passage of time. Nor is there some inviolable thing called distance or separation. Indeed he showed that the cosmos does not consist of any sort of fixed dimensions. Rather, the universe is sizeless.

These relativistic effects mean that muons, which might at first seem harmless, *can* actually harm us, and are indeed sometimes regarded as a kind of cosmic ray themselves. So along with the primary cosmic rays that create them, muons, too, belong on our list of invisible entities that continuously penetrate our bodies—and can sometimes kill us.

CHAPTER 20

Beams from the Universe's Birth

Some forms of invisible light are truly omnipresent. Satellite transmissions, TV stations, and cell towers continuously flood your bedroom with radio waves and microwaves. But before these technologies existed, even in the era of the Neanderthals, every cave dweller with furry underwear was *still* bombarded with nonstop microwaves, even if we didn't realize this until 1964. This is the natural "light" that arrives equally from all directions. It is the leftover energy from the creation of the universe.

The story of its discovery is the tale of how we stumbled upon one of the most momentous, glittery—and least disputable—keys to the nature of the cosmos.

It requires a rewind all the way back to the mid-nineteenth century, to a little-known Edgar Allan Poe essay called "Eureka." The man renowned for his poem "The Raven" muttered "nevermore" to the prevailing belief that the universe had existed more or less eternally in the same unvarying state. Poe mused that the cosmos may have begun as a sort of superdense "egg" that then exploded and expanded.

The idea competed for decades with the "steady-state" notion of a universe that pretty much always looks the same. Even after

Edwin Hubble showed in 1929 that the universe is expanding, which does suggest an explosive genesis event, this still didn't kill the steady-state idea. After all, if a single new atom of hydrogen pops out of nothingness once per century in every parcel of space the volume of a football stadium, that would replenish the empty space created by expanding outrushing galaxies. It could supply the eventual material for new stars. Such a slow, constant genesis would be very hard to detect or disprove.

So which was it? To some, including British theorist Fred Hoyle, the notion of the entire cosmos popping violently out of nothing one Saturday morning seemed preposterous. In 1949 he pejoratively referred to the absurd idea of a "big bang," and the term stuck, but without Hoyle's negative connotation. Meanwhile, who was to say which was more far-fetched—a universe that appears suddenly all at once or one that drips into existence atom by atom, like a leaky faucet? Either way, we're apparently getting something from nothing, or perhaps from an unknown dimension. Both ideas were presented in 1950s textbooks as equally plausible alternative accounts of our universe's genesis.

In the 1940s, scientists such as George Gamow calculated that if everything did begin suddenly, that blinding initial energy should still be around. True, the expansion of space would stretch out all the energy waves, redshifting the brilliance. But this invisible leftover energy ought to still be detectable. He figured that this cosmic background noise, this relic of the big bang, would produce an energy equivalent of twenty-five degrees above absolute zero on the Celsius scale.

Back then we knew neither the true rate at which the cosmos was expanding nor its actual size, and that made the exact "invisible light" frequency impossible to gauge with any precision.

Nonetheless the idea was sound, and other physicists, including those in the Soviet Union, came up with various frequencies and equivalent temperature energies that ought to be filling all of space.

By the 1960s, the consensus was that such an energy should only be a few degrees above absolute zero, the point where atomic motion stops entirely, which corresponds to wimpy electromagnetic waves only a few millimeters from crest to crest with a frequency of a few billion waves per second. In other words, they'd be microwaves. Some considered building a special radio telescope that might pick up these invisible rays, these leftovers from the beginning of time.

Everything changed on May 20, 1964. That's when two radio astronomers, Robert Wilson and Arno Penzias, who were working for Bell Labs, discovered, completely by accident, a form of radiation known as the cosmic microwave background (CMB).

The two men had been asked to help calibrate Bell's large horn-shaped radio antenna in Holmdel, New Jersey, designed for future satellite use, which at that time meant radio signals reflected off the large balloonlike *Echo 1* satellite. After all, those were the opening years of what was to be a revolution in orbiting satellite technology: *Sputnik 1* had been launched just six and a half years earlier. So they were working on baselines and calibrating against what should have been a background of outer-space silence. But they couldn't get rid of an annoying hum.

When they turned the directional antenna, the odd buzz remained. When the sun went down, and when the galaxy's center paraded through the sky, the hum persisted, as loud as before.

When TV stations went off the air after midnight, the hum continued. It maintained exactly the same volume at all times.

The scientists had no idea what they were experiencing. Maybe it was thermal radiation—heat—perhaps caused by some pigeons they'd found nesting in the antenna's metal crevices. They had an exterminator humanely remove the birds and release them a hundred miles away. Yes, the birds—the same ones—eventually returned, but in the meantime, the buzz remained; getting rid of the birds didn't affect it at all.

The physicists had no idea that just thirty-seven miles away, in that very same state of New Jersey, in Princeton, theoretical physicists were expecting exactly that kind of microwave hum to be emanating from the sky but were unaware of any existing radio telescope that might detect it!

That was the situation during the spring of 1964. One pair of physicists had discovered the buzz from the big bang but had no idea what it was. A nearby group knew that there *should* be a microwave hum if the big bang was real but didn't think there was any way to find it.

Then came an amazing coincidence. By chance, on an airline trip, Arno Penzias was seated next to Bernard Burke, a radio astronomer working at the Department of Terrestrial Magnetism, in Washington, DC, who knew about the Princeton theoretical work. Hearing Penzias describe his problematic hum, Burke urged him to phone Bob Dicke at Princeton University.

When Penzias did, Dicke got off the phone and said to his colleagues, "Boys, we've been scooped!"

So it was that Wilson and Penzias discovered the ancient

light that began saturating the universe soon after its creation. It was as accidental as finding a hundred-dollar bill in an old jacket pocket. It required no act of genius. But it was so momentous that both men were awarded the 1978 Nobel Prize in Physics. It was arguably the most easily earned Nobel in history. (Perhaps it was Burke who ought to have landed the Nobel!)

The milestone microwave discovery put the big bang theory on solid ground. It was powerful evidence that the universe did begin as a tiny sphere smaller than a baseball some 13.8 billion years ago.

Here's what apparently happened. For reasons that are mysterious and are likely to remain so, the cosmos emerged as a pinpoint from apparent nothingness, then inflated far faster than the speed of light. This inflation lasted for only a fraction of a second, then the cosmos started coasting outward on its own momentum. Conditions were so unimaginably hot that everything was pure energy. The first subatomic particles (neutrinos) formed within this energy matrix just one second later, while other particles required several minutes. Meanwhile the whole thing kept explosively growing in size. For the next 377,000 years, all forms of light, visible and invisible, were unable to travel through the opaque foggy soup of particles, which would instantly absorb and then reradiate the energy. But suddenly, 377,000 years after the big bang—a momentous moment that perhaps deserves its own holiday—conditions sufficiently cooled to allow protons to capture electrons. Neutral hydrogen atoms were born everywhere at once. At this moment, space became transparent for the first time.

Suddenly the fog lifted. The dazzling light that pervaded

everything was free to travel across the universe. It is this light that we see when we detect the cosmic microwave background.

CMB would be blinding if the universe weren't expanding. Indeed it would be so intense and energetic that it would sterilize everything and prevent planets and life forms from existing. But because space itself continuously grows in size, all the energy waves traversing it are continuously stretched out. And as we know, the longer a wave gets, the lower its energy becomes.

When the universe was half its current age, it was twice as hot as it is now. Back then, CMB, though invisible, consisted of shorter, more powerful energies that had faster frequencies. It still wasn't the original, lethally blistering soup of gamma rays and X-rays, but it was a very different background from the benign radio frequencies we detect today. Back then, the background would have been classified as infrared rays.

Today the waves have been stretched further so that they're now wimpy microwaves. Precise measurements of the CMB must be performed above the earth's atmosphere to prevent their absorption by the air. These measurements were first made by the COBE (Cosmic Background Explorer) satellite, then by a higher-sensitivity microwave-measuring satellite called WMAP (Wilkinson Microwave Anisotropy Probe), which stopped functioning in 2010. The measurements are vital to understanding how the universe evolved.

They show that the CMB has a smooth thermal "black body spectrum," the kind that would be emitted by an object shedding heat evenly. The energy radiates at an effective temperature of 2.72548 degrees Kelvin (equal to 2.73 degrees Celsius above absolute zero). Or, if you prefer, this all-pervasive glow measures minus 455 degrees Fahrenheit.

This radiance has a peak emission of 160.2 gigahertz, in the microwave part of the spectrum. If expressed as wavelength, the waves are 1.871 millimeters apart, approximately the width of apple seeds.

The glow is extremely uniform in all directions. This isotropy, or smoothness, is exactly what we should observe if the energy came from a big bang that initially inflated at faster than light speed. But the evenness is not quite perfect. Interestingly, there are tiny residual energy variations of around one part in eighty thousand from each little section of the sky to the next. This *anisotropy* between each little section of sky and the next varies with the size of the region examined but remains small everywhere. The simplest and most logical explanation is that these density differences in the early universe were globs of energy, then matter, that eventually formed structures and that we see today as stars and galaxies.

The microwaves make the big bang very plausible and the steady-state theory highly improbable. Of course the ultimate reason for the big bang, or even an accurate (or approximate) description of it, is unknown. A universe popping out of empty space offers no hint of any antecedent conditions, no clue of what may have precipitated such an astonishing occurrence. Thus the all-pervasive invisible background microwaves provide some clarity about the enormous mystery of cosmic genesis—while presenting an equally big enigma at which we can only shrug in wonder.

CHAPTER 21

Energy from Our Minds

Even if it takes us on a temporary detour from mainstream science, we can't ignore a popular, far-out aspect of unseen rays and waves: energy transmitted from the brain of one person to another.

Most people believe that the brain emits waves. Many suspect that if there's such a thing as ESP (extrasensory perception), it's because our minds transmit in some way. So let's start by confirming: are there waves coming out of our brains?

In a way, yes. If we placed electrodes on your skull, they would detect regular rhythms, or pulsations, that vary in several ways. Even better, these electrical rhythms correspond with what your brain is doing!

Just as electromagnetic waves have discrete frequencies, so do brain waves. They've been mapped and named. Unlike radio waves and infrared waves, which vibrate millions and billions of times a second, the brain's repetitive patterns are much slower. They're caused by neurons firing and simultaneously pulsing in unison at the relatively sluggish pace of a few dozen to a hundred times a second.

The main brain waves are called:

- beta waves (13–38 Hz), which are generated when the brain is active—i.e., during problem solving, conversation, and so on;
- gamma waves (39–100 Hz), also emitted when the brain is active;
- alpha waves (8–13 Hz), observed during relaxation;
- delta waves (very slow, fewer than four per second), which occur during sleep; and
- theta waves (4–7 Hz), which are associated with sleep as well but are also observed during deep relaxation and meditation.

All these waves are weak by any measure. Whereas the earth's magnetic field can barely swivel a compass needle, brain signals are a billionth of that strength.

But the term *brain waves* is a bit misleading because it makes people think of the brain as a kind of radio transmitter. In reality, human brain waves are not really electromagnetic waves at all—the kind transmitted from one place to another, as infrared radiation and microwaves are. Rather, they are changes in the electrical field. They are pulses, or rhythms, in the brain's electrical functioning—patterns that very much correlate with forms of mental activity. The electrodes attached to your skull can measure that electrical field, and the "waves" are just an interpretation of the patterns observed when these weak electrical pulses are plotted. But no one has ever detected waves being transmitted from one mind to another. Indeed, nothing ever travels from a brain to a point outside it.

When measured atop the skull, the changes in the skin's electrical field, caused by neuron firings an inch or more below it, are minuscule. Think of the waves as a reflection of the ability of an electrical force to do work—to budge a charged particle, say—rather than as forces radiating into space. However, when researchers use special helmets and signal-boosting techniques such as chilling the sensors to a point near absolute zero, and when they work in a metal-clad room where all extraneous electromagnetic activity is kept outside, electrical disturbances, visualized as wave patterns, have been detected a few inches away from the sensors. But again, they're electrical firings, not electromagnetic transmissions.

Discussing invisible energy emanating from our brains can quickly take us into quackery and New Age supposition. Actual laboratory studies looking for ESP, or thought transference, have been carried out for decades and are consistently disappointing in that they routinely come up with negative results. Some such experiments have been performed at prestigious universities; others have been wacky. One of the most famous was the effort by astronaut Edgar Mitchell during the *Apollo 14* mission, in 1971.

Mitchell, who died at the age of eighty-five in 2016, did a very strange thing that was not authorized by anyone in mission control. By prearrangement with four people back on earth, he used his own leisure "off time" during the mission to perform ESP experiments involving a series of twenty-five cards printed with Zener symbols—a circle, a cross, a set of waves, a square, and a star.* The idea was to see if mental telepathy could work

* The symbols were named for twentieth-century American psychologist Karl Zener, who used them in his experiments involving perception.

between the earth and another celestial body. The result? This depends on whom you ask.

The idea was for subjects to choose the correct sequence of cards Mitchell was looking at. Because forty correct hits out of two hundred tries would be expected by chance, and three of the four subjects scored lower than this—only the final subject scored higher, with fifty-one out of two hundred correct—no objective scientist would declare this anything but a random result, a failure to demonstrate ESP. Moreover, the launch's forty-minute delay meant that all four subjects, who had been coached to expect ESP transmissions at a specific time during the mission but who were unaware of the delay, were attempting to receive Mitchell's thoughts at the wrong time, when he wasn't even "sending" them.

None of that seemed to faze Mitchell, who devoted much of his post-*Apollo* life to psychic research. In his analysis, he characterized the results of his ESP experiment as "statistically highly significant." Indeed, he eventually explained that "the well-known experiment in the laboratory was to use cards with the five Zener symbols, but the actual cards aren't important. It was easier for me to use random number tables than carry the physical cards. Instead, all I did was to generate four tables of 25 random numbers just using the numbers 1 to 5. Then I randomly assigned a Zener symbol to each number. For each transmission, I would then check the particular table of random numbers and think about the corresponding symbol for 15 seconds. Each transmission took about 6 minutes." At any rate, Mitchell kept the entire project secret from both NASA and his crewmates. He later explained that NASA's feelings about such psychic research was "totally negative, totally closed"—an appraisal that surprised no one.

* * *

Science not only fails to provide empirical support for the theory of extrasensory perception, it also offers an excellent explanation for anecdotal ESP events that do seem to work. The reasoning goes like this:

Say you're thinking of a particular old friend, and just then the phone rings, and it's the friend you were thinking of. You might ascribe it to telepathy and share that with her ("You read my mind!"). We all notice and remember whenever such a remarkable coincidence happens because it's so dramatic. Yet how many times do we idly think about lots of other people who do *not* call at that moment? We don't even give those nonevents a second thought, let alone remember them. Thus we experience *bias selection*—that is, we recall only those times when ESP is supposedly at play.

Bias selection is a good explanation for the apparent instances of ESP most of us have experienced in life. It's hard to rebut. Yet according to Gallup polls, between 50 and 60 percent of respondents think that "some people possess psychic powers or ESP." How do we account for that? Is it mere gullibility? Even dismissing the stubborn minority who are chronically naive or antiscience (in 2001, 30 percent of National Science Foundation survey respondents thought that "some of the unidentified flying objects that have been reported are really space vehicles from other civilizations"), we're still left with a substantial number of people who think that mind reading is real.

In truth, while all manner of New Age psychic nonsense has been adequately discredited, there still remains a possibility that our minds really can interact with those of others. Perhaps we are linked more deeply than we know: the mysterious workings of the brain might be capable of connections that transcend today's known modalities. Here are a few examples that, for me, at least, make it hard to completely rule out this possibility.

In the first instance, I was playing Scrabble with a friend, waiting for her to finish her turn. Staring idly at the board, I suddenly and vividly saw, in my mind's eye, her hand coming down and placing an *I* in front of an existing *T* on the board to make the two-point word *it*.

A couple of seconds later, she did exactly that—even though it was a most unlikely play. (What Scrabble player would ever squander a turn for a measly two points?) Startled, I asked her if she had *visualized* making that word a few seconds before she actually did it, and she said yes. So I'd apparently seen it in my mind at the same time as she did. Advance perception of such an unlikely event in two different people, at the exact same moment in time, seems too improbable to dismiss as bias selection.

In the second instance, two close friends of mine who are identical twins are both seriously into music and singing; each was in his college's choir. Throughout their lives, they both insist, one twin would have a song playing in his head when suddenly the other twin would start to sing it, beginning in the exact same place in the tune *and in the same key!*

Another identical twin told me an interesting story about something that happened when she and her sister were teenagers: "When we were seventeen, my sister went to live in Mexico for six months while I went to live in the South of France. When we got back we figured out that each of us had translated a song from the cartoon movie *An American Tail*—"There Are No Cats in America"—into the local language, then had it stuck in our heads for months."

The coincidence explanation sounds unconvincing in cases

like these. Obviously some satisfactory mechanism must be found to explain such strangely synchronous human events.

Okay, I know what you're thinking (ha!). Those were identical twins, who may live in a world apart from the rest of us. But still. I will ask skeptical readers to forgive me here, but I simply cannot fully dismiss the possibility of ESP, even if it seems incapable of replication in controlled laboratory settings. Anyway, if it is real, the explanation may indeed involve certain brain waves that can march in sync with others. Because electrical pulses are always accompanied by electromagnetic wave activity (recall that Hertz found radio waves by noticing sparks across gaps), perhaps ESP also works along these lines.

Belief in ESP appears to have remained steady over the decades. According to a 2002 CBS News poll, a majority of Americans—57 percent—believe in ESP, or telepathy. But despite this, only a small number say they've actually had such an experience themselves. Some interesting demographic differences also came to light. Americans age sixty-five and older are most skeptical about ESP, telepathy, and other similar experiences. Thirty-two to 47 percent of people in that age group do not believe in it. But 31 to 67 percent of people younger than sixty-five think that people can sometimes perceive the thoughts of another. Interestingly, those with at least some college education are *more* likely to believe in ESP.

Except, perhaps, for the discussion about the effects of microwaves on human health, this must be the only chapter in this book in which I allow that evidence concerning the putative consequences of invisible rays is inconclusive. When it comes to the mind's unseen energies, you are not discouraged from healthy skepticism.

CHAPTER 22

Ray Guns

Deadly beams from handheld or spaceship-mounted armaments have been sci-fi staples since forever. What *Star Wars* or *Star Trek* movie would be complete without directed-energy weapons—things like laser cannons and tractor beams?

The idea of an energy ray used in warfare probably dates to its supposed employment by the ancient Greeks. The singular event was a dramatic battle between the Romans and Greeks in which parabolic mirrors supposedly set fire to invading ships. So before we explore ray guns and other invisible-beam weapons—some of which actually now exist—we must start with the father of all such ideas, Archimedes, and his defensive tactic, conceived in 212 BCE, of burning approaching enemy ships with focused energy.

Most people know the gist of the story: Archimedes set fire to Roman ships attacking his home city of Syracuse by focusing sunlight. But to this day there's controversy about the weapon he used. Did the brilliant Greek inventor employ a giant concave mirror? Or did he deploy a Greek battalion wielding hundreds of mirrors?

You'd think historians and researchers would have settled the matter long ago. Books have been written about the Archimedes heat ray; experiments trying to replicate it have been performed, and the US TV show *MythBusters* tested the problem in two separate episodes, most recently in 2011.

Unfortunately, the earliest surviving accounts of the event date from 1,400 years after the battle—a long interval indeed. Byzantine scholars Joannes Zonaras and John Tzetzes, writing in the twelfth century, paraphrased parts of a work by Cassius Dio called *Roman History,* published in the third century CE. The section about the siege of Syracuse said, in part:

> When Marcellus [the Roman general] had placed the ships a bow shot off, the old man [Archimedes] constructed a sort of hexagonal mirror. He placed at proper distances from the mirror other smaller mirrors of the same kind, which were moved by means of their hinges and certain plates of metal. He placed it amid the rays of the sun at noon... [and] a frightful fiery kindling was excited on the ships, and it reduced them to ashes, from the distance of a bow shot.

The original source of this account survives only in fragments. So the most logical way to verify this fascinating event in our modern techno-savvy age would be to construct the kind of mirror Archimedes might have created and see if it works.

The method seems reasonable. Most of us, as children, focused sunlight using a magnifying lens and burned a hole in a piece of paper. We discovered that the color of the paper made a huge difference. White paper let the dazzling sunbeam linger a long time with no initial effect. Smoke would eventually appear, and all the while the tiny focused spot of sunlight was uncomfortably brilliant to the eye. Black paper, by contrast, catches fire fairly quickly. So right from the start we see an obstacle for Archimedes: the enemy Romans were unlikely to have obliged him by using dark sails, which would have helpfully absorbed rather than reflected the heat-producing infrared radiation from his mirror.

One can also concentrate sunlight (and, as we'll soon see, invisible rays) using a parabolic mirror instead of a lens. Indeed, a mirror is much easier. Any hobbyist with an old telescope mirror can ignite a fire within seconds, even if the mirror is just six inches across. The distance from which the fire can be set depends on the mirror's focal length. Telescope optics bring rays to a focus a few feet away, but if a telescope was built to achieve focus at a few hundred yards, it could easily do the job Archimedes intended.

Current technology uses mirrors to concentrate the sun as a way of generating electricity. A huge solar array of two thousand swiveling and tracking mirrors in the Mojave Desert near Barstow, California, focuses sunlight on a tower, where molten salt absorbs the heat and transfers it to a water boiler below. The resulting steam turns generators that produce ten megawatts of electricity. A giant array like that could ignite wood or sails virtually instantaneously.

One hurdle for Archimedes is that every mirror has a fixed focal length. How could he know exactly how far away the Roman ships would be? If you misjudge the distance, the light is spread out and you can't achieve the kindling temperature for sails, which is around five hundred degrees Fahrenheit. Moreover, it would take an enormous mirror to do the job, one at least ten feet across. On top of that, the metal mirrors available to Archimedes in his day would only reflect around 65 percent of the light hitting them, compared with the 90 percent that today's "silvered" mirrors reflect. Size for size, Archimedes would have needed three old-fashioned mirrors to do the job of two modern mirrors.

Practically speaking, it appears that success would be most probable if Archimedes had a large number of his soldiers each hold separate flat mirrors rather than try to guess the correct distance so that he could fashion a giant parabolic mirror of the required shape. If the soldiers all stood aligned in an arc, they could create a crude parabola, or they could simply stand anywhere they wanted and aim their spot at a single place on the enemy ship.

Testing this idea in 1973, a Greek scientist named Ioannis Sakkas had sixty Greek sailors each use a three-by-five-foot flat mirror to focus sunlight on a wooden rowboat 160 feet away. He reportedly succeeded in starting a fire "very quickly."

But hold on. In 2009, a class at MIT tried to reproduce Archimedes's feat using 127 one-by-one-foot mirrors. They did get a large piece of dry red oak (representing boat material) to start burning after ten minutes—and that was using mirrors with just one-seventh the total area of the 1973 experiment. A later MIT experiment used an actual old boat and achieved similar results. But ten minutes is a long time to stand there in the heat of battle—especially if the sailors on the boat were meanwhile tossing pails of water on the bright spot the moment it started smoldering.

MythBusters tried to settle the matter in an episode aired in 2010, with the help of volunteer middle school and high school students holding five hundred mirrors. Their results were discouraging. Despite an hour of focusing the sun on a sail, a ship's weak link when it comes to flammability, they could only get the temperature of the beam up to around 230 degrees. That would have boiled water but not initiated combustion.

Adding to the discouragement is the fact that writers of Archimedes's era, including the famous Plutarch, never mentioned any use of mirrors to set ships on fire, even though they amply described other innovative devices used by Archimedes. Centuries later a few writers did say that Archimedes set fire to the Roman ships, but they omitted any mention of mirrors, which should have been dramatic and unusual enough to merit inclusion in the story. Most analysts assume that Archimedes used other fire-starting measures, such as petroleum. (Ultimately his strategies failed. He was killed by a Roman soldier during the ensuing battle, even though the Romans had standing orders to spare his life when they found him.)

The consensus is that the story is probably apocryphal, although not impossible. Still, its fame has kept alive the idea of a weapon that focuses energy.

Fast-forward to 1898, when H. G. Wells introduced the fearsome Martian weapon used against Earth—the Heat-Ray. William Herschel's old term *calorific ray* had already changed to *infrared ray* decades before *The War of the Worlds* became a bestseller. But "Heat-Ray" sounds cooler and scarier than "infrared ray." In 1953, in John W. Campbell's sci-fi bestseller *The Black Star Passes,* we find the first-ever mention of something called a ray gun. But gadgets that shoot "disruptor rays" and "annihilator beams" appeared even earlier, in stories from the 1930s, and became a standard armament in the Buck Rogers stories. (In the 1951 movie *The Day the Earth Stood Still,* a lethal ray disintegrates only inanimate objects, a humane characteristic seen nowhere else in fact or fiction.) The idea behind a "disintegrator beam" was that it somehow nullifies the electrical forces that bind the subatomic particles within atoms, leaving the material to fall apart as ionized fragments. In reality, to make atoms disintegrate, a weapon would also have to nullify the strong force that keeps nuclei intact—but no sci-fi author seems to have bothered with such particulars.

While writers and filmmakers dreamed up hypothetical directed-energy weapons, the technology that could enable them was close at hand. In 1957, Columbia University graduate student Gordon Gould figured out a way to make photons march in unison, a phenomenon predicted half a century earlier by Albert Einstein. He even coined a term for it: LASER—light amplification by stimulated emission of radiation. A laser could create and focus coherent streams of visible or invisible light with equal ease. At this time, Bell Labs was also furiously trying to find a way to make light waves' peaks and troughs pulse in unison. When they built the first usable laser, in 1960, there ensued a patent fight that wasn't resolved for seventeen years. Historians still debate the matter of who really invented the laser.

Nobody foresaw how quickly this invention would enter our daily lives and how ubiquitous it would become. It changed everything from how we shop for groceries to how we listen to music. Nor could anybody have predicted how inexpensive it would become—and how quickly it would do so. In a mere fourteen years, the Universal Product Code, with its black bars separated by varying spaces, which convert a laser reflection into on-and-off signals that produce a twelve-digit numerical code readable by a computer, had been created. The National Cash Register Company installed a test bar-code system at the Marsh Supermarket in Troy, Ohio, near its factory producing the equipment. On June 26, 1974, at 8:01 a.m., shopper Clyde Dawson pulled a ten-pack of Wrigley's Juicy Fruit gum out of his basket, and it was scanned by cashier Sharon Buchanan. This was the first commercial use of the UPC. The pack of gum and the receipt are now on display in Washington's Smithsonian Institution.

Supermarket lasers, like those in CD players, use around five milliwatts (mw) of energy, which is also the legal limit for handheld laser devices such as presentation pointers. Lasers in DVD players use up to 10 mw, and DVD burners require 100 mw. In all these cases, laser reflections from the spinning disk's tiny flat areas, called lands, combined with the absence of reflection from deeper "pits," produces a rapid succession of laser pulses, a code that the built-in computer chip converts to digital information that is then translated into images and sounds. Lasers used in surgery, on the other hand, don't create coded information. Instead the goal is simply to use its sharply concentrated heat. Laser scalpels employ 30,000 to 100,000 mw of energy, meaning 30 to 100 watts, to effortlessly cut through flesh.

Everyday five-milliwat red lasers used for pointers and cat toys are the least expensive, at a few dollars apiece. These, however, do not create a visible beam in the night sky. That's because their emissions can't illuminate airborne dust or tiny water drops that in turn reflect light back to the user's eyes.

To produce a visible ray, you need a green laser or one of the newer blue or violet lasers that allow reflection off airborne particulates such as pollen and dust. Because green is perceived far more readily than any other color, it's the only one that can create a visible beam using just the legal 5 mw of energy. ("Legal" because it's the maximum output the government allows for a handheld laser. As of 2017, however, more powerful devices can still be purchased online.) But in bright moonlight or in light-polluted cities, a 30 mw or higher green laser is the best way to create a nice visible beam. Available from specialty websites, many in China, they inhabit a quasilegal niche. But even the unambiguously legal 5 mw lasers are inherently dangerous if pointed at an eye. Laser light is so concentrated that it can produce eye damage in a fraction of a second.

Since 2007 or so, an increasing number of young people are buying superboosted 20 mw, 50 mw, and even 100 mw lasers. These are fabulous tools for showing off constellations, and companies such as Wicked Lasers, exploiting the popularity of the lightsaber battles seen in *Star Wars* movies, produce ever-more-powerful models with teen-friendly names like Spyder and Krypton. I've tried several of these. One was a 1,000 mw model—a full watt! Point one of these at a dark-color balloon, and it will pop almost instantly. But their reflections off glass or chrome surfaces are dangerous to look at. These handheld devices are available on the Internet for $300 or so.

At night, without much thought to the consequences, some people point them at passing planes. Inside the cockpit, pilots are suddenly incapacitated—totally blinded for several seconds. Sometimes the pilot cannot continue his or her duties because the resulting headache and dizziness persist for hours. No one has yet been permanently blinded, but with lasers getting increasingly powerful, it may only be a matter of time before this happens.

In 2012, President Obama established a penalty of five years in prison for the offense of pointing a laser at a plane. Soon afterward, one California man received a sentence of fourteen years because of his "willfulness" in repeatedly zapping helicopters.

Scanning bar codes and playing at lightsabers is all very well. But what about actual, real-life directed-energy weapons? They've gone from sci-fi fantasy to reality. The US military continues to develop ultrapowerful laser weapons using the invisible sections of the spectrum.

During World War II, British scientists tried to focus microwaves—which were already successfully used in radar and would later be used in microwave ovens—into a sort of death-ray gun, but it never panned out.

A version of a microwave-beam weapon appears in a 2006 sci-fi book, *Blindsight* by Peter Watts, along with the popular misconception that microwaves cook flesh from the inside out. (Microwave ovens cook food all at once—the inside and outside simultaneously, albeit with some unevenness—see page 110.) But *actual* focused microwave weapons have indeed been developed and deployed by the US military. The army version, which uses a parabolic dish on a Jeep-like vehicle, can create heat in organic targets, but this has been fairly ineffective in practice. Sometimes target victims have reported a sensation of increased warmth. Despite much publicity about expensive but failed laser antiballistic missile systems, more modest systems have become a reality.

In 2014, LaWS (the Laser Weapon System) went into operational status aboard a US naval vessel. This ship-defense system uses a thirty-thousand-watt infrared laser from a solid-state array and has successfully shot down and crippled incoming test targets. It was also able to render a small approaching boat's engines inoperative.

Even the most powerful lasers, which can quickly melt holes in metal objects, are limited in their effectiveness by atmospheric scattering. They also have difficulty keeping a quickly moving target, such as a missile, precisely centered in their sights. Nonetheless, current truck-mounted high-energy laser cannons can reportedly shoot down aircraft.

The latest lasers supposedly use powerful magnetic fields to boost the speed of electrons, which then transfer energy to the laser's photon beams. As of 2016, these are too bulky to be employed as handheld or even truck-mounted weapons, but research and development continue.

Now that we know how invisible-beam armaments have progressed, it's easy to answer those who imagine there are effective current ray-gun weapons in the world's military arsenal. On the Web, some people paranoiacally argue that the smoke and dust that radiated from the World Trade Center buildings as they collapsed is proof that they were destroyed by secret US military ray guns. When an educated friend actually said that to me, I was mystified.

"What part of the electromagnetic spectrum would this ray gun be using?" I asked. An energy ray is made up of electromagnetism; there is no other kind of focusable energy except a sonic or sound-producing device. It takes only a few moments to consider the entire electromagnetic spectrum. Such a weapon couldn't use gamma rays or X-rays—these don't harm concrete and metal. Nor could it use ultraviolet rays, which would similarly cause no damage at all. It couldn't use visible light, because that's harmless, too—not to mention that no sudden brightness was observed just before the buildings collapsed.

A ray gun couldn't use radio waves, including microwaves, because no matter how intense they are they cannot make concrete disintegrate. That leaves infrared radiation—heat. A theoretical unlimited-power focused heat ray could indeed melt steel. But any building shot by such a ray would first glow red, then white, before it melted. And it would radiate enough heat to cook all the pedestrians on the surrounding streets. Bottom line: no invisible-electromagnetic-ray weapon could cause buildings to collapse into smoke and dust—all without creating any heat.

Ray guns remain in the realm of sci-fi for now. But knowledge of invisible light helps us understand the nature of all invisible beams, including potential weapons. Thus far, directed-energy weapons sound a lot scarier than they've proved to be in actuality.

And FYI: if you were attacked by any kind of light-beam weapon, whether it uses visible light ("photon torpedoes") or invisible rays ("microwave cannons"), you'd never see the beam approaching. So those movies in which our hero weaves and ducks his spacecraft to avoid incoming photon torpedoes—forget it. You would have no advance notice of something approaching at light speed. You couldn't see a light beam or a photon torpedo (an intense cluster of energy) before it arrives, because its image *is* the weapon. Just a little something to keep in mind the next time you watch *Star Wars*.

CHAPTER 23

The Next Frontier: Zero-Point and Dark Energies

Our quest has been to explore the unseen energies that pervade our universe, our planet, and our bodies. We've seen that the word *energy* usually refers to waves on the electromagnetic spectrum. But since 1948, science has slowly awakened to an unexpected reality: some other superpower not only lurks everywhere, it may also dwarf all other energies combined.

Ah, for the good old days, when *vacuum* meant "nothing."

Those days are gone forever. We now have reason to believe that the universe's vast tracts of emptiness seethe with unimaginable power.

In a way, it all starts with the ancient Greeks, who hated the notion of a vacuum. Their argument was semantic, stemming from their love of logic. How could there be a vacuum, they reasoned, when the words *be* and *vacuum* are contradictory? If a vacuum is nothingness, well, there can't *be* nothing.

Those silly Greeks, we all thought back in the 1960s, when I first studied physics in college. Of course you can create a vacuum. Just evacuate all the air out of a bell jar, every last molecule, and voilà. Our laboratory vacuums still had a few molecules in them; they weren't perfect vacuums, but so what? That hardly mattered, because the basic premise, which is that nothingness is real, seemingly remained valid.

Turns out that the Greeks were right. First, no matter how good the vacuum, its space is still penetrated by some heat, in the form of infrared radiation and microwaves, emanating from the vacuum's walls or its environment. Because energy and mass are fundamentally the same, those waves zipping through all of space ensure that you can't ever have a true vacuum.

But that's small potatoes compared to Werner Heisenberg's *uncertainty principle,* which says that a vacuum shouldn't actually exist. He was quickly followed by other theorists who argued that the vacuum of space should be filled with a bizarre sort of quantum energy.

They were right, too. Much experimental evidence shows that virtual particles—things like electrons and antimatter positrons, which are electrons with their electrical charges reversed—snap, crackle, and pop out of nothingness everywhere all the time. Each particle typically exists for just a billionth of a trillionth of a second, then vanishes. If there's an energy field around, a subatomic particle can use some of the energy to remain in existence forever. Thus things perpetually spring to life out of that quantum vacuum.

Most physicists now believe that this underlying "quantum foam" pervades the cosmos. It's everywhere, and its power is unimaginable. Estimates of the power in each small bit of seemingly empty space vary enormously. But it's possible, perhaps likely, that the apparent nothingness within an empty coffee cup held by an astronaut in the virtual vacuum of space contains enough energy to instantly boil away all the earth's oceans.

Wait a second. I'm from Missouri. Prove it to me.

Okay, consider the Casimir effect. It was named for the Dutch physicist who first predicted it in 1948. He said that if you suspend two copper plates very close to each other—a millionth of an inch apart—the quantum energy between the plates would suffer a limitation: its waves wouldn't have enough space in which to flow, just as microwaves are physically too big to fit through those little holes in the screen in your microwave oven's door. But the quantum energy *outside* the two plates would be as strong as ever, so it would push the plates together—hard. Well, this really happens. The Casimir effect is real.

Some dreamers think of exploiting "vacuum energy" to give the world unlimited power. But there's a problem. This energy exists everywhere equally, which is why we don't sense it or detect it. Energy only flows from a place of greater energy to a place of lesser energy. So how would you set up a condition that had less energy than everything around you? How could you make it come to you?

The only way to do that is to chill matter to absolute zero, at minus 459.67 degrees Fahrenheit, the point where all molecular motion stops. Then and only then are things at parity with this all-pervasive power, which is why it's also called zero-point energy. That's where this hidden fount of energy becomes evident: why else would helium still be liquid at absolute zero if it weren't receiving a bit of energy that keeps it from freezing solid?

In short, zero-point energy does show itself when all other energy is absent.

But in order to get this limitless quantum-foam energy to flow to you, you'd have to somehow create below-absolute-zero conditions. You'd have to make molecules move more slowly than "stopped."

More slowly than stopped? How do you do that?

If you have any ideas, we're all ears.

In the meantime, imagine that the seeming emptiness of space is not just permeated by unseen particles such as neutrinos, unseen magnetic and electric fields, and unseen waves of microwave energy, infrared radiation, and the like. It is also permeated by unimaginable power. Because we're barely in our infancy when it comes to recognizing the existence of zero-point energy, we can't say what it may or may not do to us. For the moment, visualize what you *can* see as being a thin, almost inconsequential surface scum that overlies this lively three-dimensional energy.

We can't see or feel all the invisible energy that surrounds us because there'd be no biological advantage to that. Why perceive a blindingly powerful energy that's equally present everywhere? Perhaps if we could fast-forward to a point in time one hundred years from now, we'd find that our technology's obliviousness to this zero-point energy seems as primitive as the attitude of the eighteenth-century scientists who didn't lift a finger to try to exploit electricity.

Still, file zero-point energy away somewhere under "most powerful entity of all." Surely as technology and our understanding of it improve, we humans will not ignore vacuum energy forever.

Beyond zero-point energy, there are other mysterious, newly discovered energies in the universe. The existence of vacuum energy had been theorized in the 1930s, and solid evidence for it started to emerge in 1948. But a much more recent development uncovered the apparent existence of something else entirely, which we call *dark energy*. Until 1998 it was an unknown, and indeed no one had even hypothesized its existence. It was one of the most unexpected discoveries in the history of science, right up there with William Herschel stumbling upon the planet Uranus.

You see, we'd been observing the expanding universe since the late 1920s. All observations seemed to show that the cosmos started explosively expanding 13.8 billion years ago and has been steadily slowing down ever since.

Cosmic deceleration made perfect sense. Gravity from every individual galaxy pulls on all others. Logically, the outrush should be continuously diminishing, just as it does when you let go of a stretched-out rubber band. The big question was whether it would someday come to a stop. Then perhaps the universe might go the other way and get smaller. To find out whether this was indeed happening was the scientific holy grail of the 1980s and the 1990s. Everyone was looking for the *deceleration parameter*—the exact rate at which the cosmos's expansion was supposedly slowing down.

Measuring galaxies' radial velocity, their recessional speed, was the easy part; it could be done simply by using the famous Doppler redshift. But this required that we also know their exact distance from us. In the late 1990s two teams of astronomers studied a particular type of supernova, type 1a, whose brightness was believed to always have the same intensity. They used these supernovae as "standard candles" to determine the distances between the thousands of galaxies in which they erupted. When the data came in, it threw everyone for a loop. It turned out that the universe's expansion was not slowing down at all. When the smoke cleared, a new picture suddenly emerged. Unfortunately, it told a story that made no sense.

It seems that for the first half of the universe's life, its expansion did slow down. But then, around seven billion years ago, all galaxy clusters started speeding up. Since then the expansion has gotten more animated. There's no end in sight to this bewildering acceleration.

What could have made galaxy clusters suddenly move away from one another with increasing speed? Could they have had enormous rocket engines on them that fired in unison when the universe was half its present age? What could possibly be happening?

Out of desperation, physicists coined a new term—*dark energy*—to describe a kind of antigravity force that pervades the cosmos by lurking within empty space itself. Presumably, in the early eons of the universe, things were so crowded that gravity overcame the outward push of dark energy. But when the cosmos became empty enough, the antigravity property of empty space started assuming command. From that point on, the inherent repelling nature of space was enough to overcome gravity and dominate the cosmos, making everything expand like an over-yeasted bakery roll.

So dark energy fills the entirety of empty space and was perhaps the impetus for the big bang in the first place. If so, the universe is still banging. Nonetheless, nobody knows anything about what dark energy really is, except that it has antigravity power.

For those who tend to worry about things and are not content to fret about blood pressure and cholesterol, this runaway expansion may seem depressing. It seems to point to an ever-lonelier universe, with a future of ever-greater spaces yawning between galaxies. Of course, since we know nothing about dark energy, including how it arose, we can't say whether it's permanent or even whether it may someday reverse itself.

Also, while dark energy appears to dramatically show itself in large-scale phenomena, perhaps in smaller quantities it lurks everywhere. But that's about as far as we can speculate.

Suffice to say that, along with the invisible energies we know a lot about, we've now got these other recently discovered phantoms whose true natures must remain tucked away in our X-files cabinet.

CHAPTER 24

Total Solar Eclipse: When the Rays Stop

Have you ever witnessed a total solar eclipse? Usually, when I give a lecture, only a couple of people in an audience of several hundred raise their hands when I ask that question. A few others respond tentatively, saying, "I *think* I saw one." That's like a woman saying, "I *think* I once gave birth."

What these people are remembering is some long-ago *partial* solar eclipse. These are quite common. They occur every few years in various places across the globe. But believe me, if you've seen a total solar eclipse—when the moon passes directly between the sun and the earth—you'll never forget it.

Part of what makes a total eclipse so breathtaking has to do with invisible light. During the "moment of totality"—the minutes during which the sun is completely blocked—observers experience the exquisitely odd and wondrous sensation of solar emissions, both visible and invisible, vanishing right in the middle of the day.

We had our chance to experience this firsthand. The United States reached the end of the longest total-solar-eclipse drought in its history. A total solar eclipse—or totality—had not been observed from anywhere in the mainland United States since February 26, 1979. This bizarre thirty-eight-year hiatus ended on August 21, 2017, when a coast-to-coast totality swept across the continent, ramping up a much-hyped eclipse fever.

For those who did not live in or travel to the narrow, ribbon-like path of totality—the area from which the sun appeared to be in total eclipse, which stretched from the Pacific Northwest to the Carolina coast but was just 150 miles wide—a second totality will unfold on April 8, 2024. Two in a mere seven-year period.

Then, as if to compensate for the scarcity of these events (even the 1979 eclipse was a mostly cloudy, far-northern event only observable in a few places such as Helena, Montana), the middle and late parts of the twenty-first century will offer a second sudden flurry of them.

In any given place on earth, a totality appears just once every 375 years. If it's cloudy, you have to wait *another* 375 years. So a totality is a very rare event for any location. But that interval of time is just the average. Here and there, a few places will enjoy two totalities in a single decade: Carbondale, Indiana, for example, sits at the intersection of both eclipse tracks—2017's and 2024's. Yet residents of other cities, including Los Angeles, must cool their heels for more than a millennium.

In the United States, no major urban center has seen a total solar eclipse since the dual events of Southern California in 1923 and the now-famous New York City totality of 1924. Boston was scheduled for a sunrise totality in October of 1925, but it was cloudy.

Every eclipse path—a map of the places on earth from which the sun is completely blocked and where stars are seen during the day—is long and narrow. It's usually around 150 miles wide, but its length extends for thousands of miles. During that Roaring Twenties Big Apple eclipse, for example, the totality ran from central Canada southeast to Albany, in upstate New York, then down through the Bronx and Harlem, and ended unceremoniously at 86th Street in Manhattan, near an eatery that would someday be famous for hot dogs and papaya drinks. People south of the subway stop there stood in daylight: no stars out, no mind-numbing glimpse of the solar corona, no hot-pink flares shooting from the sun's edge. Volunteers were dispatched to each street so scientists could later know the precise location of the edge of the moon's shadow. The next day, a newspaper writer, watching the disappearing sun's final dazzling pinpoint, described it as a *diamond ring*—a term that has since been fully incorporated into eclipse-speak.

The event has an indescribable effect on observers. While most experienced astronomers would concede that a total solar eclipse is the most powerful, gorgeous, and even life-altering of all celestial phenomena, they'd rate a vivid display of the northern lights as not too shabby, either. A big gap separates those two from the rest of what I call the top four natural spectacles, including a rare brilliant comet and a meteor storm, in which more than a dozen shooting stars flash across the sky each minute. Like the aurora borealis, a solar totality often invokes involuntary gasps and cries of wonder. You'll often hear that some kind of "feeling" accompanies the visual spectacle. Perhaps this has to do with the fact that both these events are indeed accompanied by large changes in the amount of incoming electromagnetic radiation. It should also be noted that lunar eclipses, even total ones, do not make this top-four list. Those fairly commonplace eclipses, which unfold every few years and are never limited to a narrow section of our planet but instead are visible to half the world, are certainly pretty and worth watching. But they are not life-altering.

During a solar totality, animals usually fall silent. People howl and weep. Flames of nuclear fire visibly erupt like geysers from the sun's edge. Shimmering dark lines cover the ground.

In both the 2017 and the 2024 events, the entirety of the United States and Canada will experience a *partial* eclipse, so

The moon blocks out the entire sun except for a tiny section, the "diamond ring," during the 2012 Australian total eclipse. But no photograph can remotely capture the event's true glory. *(Matt Francis, Prescott Observatory)*

that anyone using protective eyewear will be able to see it by standing outside or by looking out a window (provided that it's not cloudy, of course). In contrast, less than 1 percent of the continent will experience totality. To most people, it might seem that seeing a partial eclipse ought to be almost as good as seeing a total eclipse, and it's certainly a lot more convenient. Why travel? The sun being 99.9 percent eclipsed doesn't sound too different from its being 100 percent eclipsed, right? Actually, seeing an *almost* total eclipse is no better than *almost* falling in love or *almost* visiting the Grand Canyon. Only full totality produces the astonishing and absolutely singular phenomenon that resembles nothing else in our lives, on our planet, or in the known universe.

No discussion of totality should omit the strange science lurking behind it. It starts with a bizarre coincidence: the moon is four hundred times smaller than the sun, but it also floats four hundred times nearer to us. This makes the two disks in our sky appear to be the same size. Now, if the moon appeared larger than the sun, it could still occasionally stand in front of it, but it would also blot out the dramatic prominences along the sun's edge, those geysers of pink nuclear flame. So for maximum amazingness, these bodies must have identical angular diameters—i.e., they must appear to be the same size. And they do.

The moon wasn't always where it is now, which makes the coincidence even more special. The moon has really just arrived at the "sweet spot." It's been departing from us ever since its creation four billion years ago, after we were whacked by a Mars-size body that sent white-hot debris arcing into the sky. Spiraling away at the rate of one and a half inches per year, the moon is only now at the correct distance from our planet to make total solar eclipses possible. In just another few hundred million years, total solar eclipses will be over forever.

For early cultures that regarded celestial phenomena as magical to begin with, eclipses occupied a spot entirely off the weirdness scale. Some, such as the Aztecs and the Babylonians, were obsessive enough to make astoundingly accurate observations that ultimately gave their priests the power to predict astronomical events.

The ancient Babylonians noticed that although some sort of eclipse happens every year, the exact same type of eclipse returns after precisely eighteen years and eleven and one-third days. The accuracy of this observation remains very impressive, especially because that one-third-of-a-day business means that the next eclipse can be best seen (or maybe *only* seen) in an entirely different region of the world. Babylonians called this eighteen-plus-year period a Saros. The ancient Greeks loved that word and concept so much that they embraced it without even translating the word into their own language.

The Saros's third-of-a-day feature means that the earth turns through 120 degrees of longitude before the next eclipse in that particular Saros takes place. Therefore, for an eclipse with specific properties (such as total versus partial, long versus short, and tropical versus arctic) to make a repeat appearance in any particular region, one has to wait while eclipses work their way around the world like a set of gears, which requires three Saroses—a length of time equal to fifty-four years and around one month, or, more precisely, thirty-three days. Because this surpasses human life expectancy in that era four thousand years ago, it's astonishing that the cycle was noticed at all. This three-Saros interval is called the exeligmos, which is Greek for "turning of the wheel." Using the exeligmos, we can calculate that there must have been a total solar eclipse in the United States fifty-four years and one month before the 2017 event and fifty-four years and one month before the 2024 event. Sure enough, a total eclipse in Maine unfolded in 1963, and another one amazed onlookers when it raced up the East Coast and covered Virginia Beach and Nantucket on March 7, 1970.

That three-and-a-half-minute March 1970 totality over Virginia Beach belongs to a series of Saroses given the number 139. This series consists of total (not partial) eclipses with paths that always move northeastward. In 1988, this Saros presented its next event a third of the world west of Virginia—a three-and-three-quarter-minute totality over Indonesia. Yet another Saros later, in March of 2006, the same northeastward totality swept from Libya to Turkey. Saros 139's next return, another third of the world west, will show residents of Cleveland, Rochester, Buffalo, and Burlington, Vermont, a totality in 2024.

So now our stage is set for the next eclipses over North America. After the two-and-a-half-minute coast-to-coast 2017 spectacle, the eclipse on April 8, 2024, will appear longest over central Mexico, at well over four minutes; then the moon's shadow will move northeastward like a tornado to the northeastern United States.

After 2017, a solar totality will happen once, somewhere in the world, during most years. None will occur in 2018, but we'll get a sunset totality over central Chile and Argentina on July 2, 2019, then another in those same countries on December 14, 2020.

Ignoring a strictly Antarctic totality in 2021 and the eclipseless year 2022 takes us to a marginal one-minute event in steamy equatorial Indonesia in 2023. But then things pick up, convenience-wise.

The 2024 US totality will be followed by the totality of the longest duration between 2017 and the end of the century—six and a half minutes—which will occur in Egypt and Gibraltar on August 2, 2027. That decade will be rounded out by a wonderful five-minute Australian totality on July 22, 2028.

If you want to limit your eclipse tourism to the United States, Canada, and Europe, note that the United States will see its longest-ever solar eclipse on August 12, 2045, a six-minute totality running from Northern California to Florida. Florida gets another eclipse just seven years later, on March 30, 2052. Then the United States will enjoy two within a twelve-month span, on May 11, 2078, and May 1, 2079, while France and Italy will experience their only totality of the century on September 3, 2081.

I have had the good fortune to see eight totalities; please allow me to share the experience. The fully eclipsed sun is always a breathtaking surprise.

First off, no one is really prepared for a total eclipse. Pictures one may have seen don't do the event justice, because cameras never capture its true visual appearance. The reason has to do with the difference between human retinal sensitivity and the vagaries of a camera's exposure, whether using digital imaging or film. The inner corona is bright; the outer corona faint and delicate. The correct exposure for one part of the eclipsed sun either underexposes the other so that it's invisible or overexposes it so that it looks like a huge burned-out area ringed by wide white flares. So a real eclipse does not resemble the ones you see on nature documentaries or in magazines, even when the images are taken by professionals. To get an accurate image, you would have to Photoshop multiple images together.

The magic really starts around ten minutes before totality, when the sun is still partially blocked but almost gone. You need eye protection at this point; I prefer welding goggles fitted with shade 12 filters if the sun is low and shade 14 if the sun is high. These display a clearer, higher-quality image than cheap plastic eclipse glasses do. (Get the goggles from a welding supply store, which is absolutely never located in the mall but rather in the worst part of town, usually adjacent to a fenced-in yard protected by snarling dogs.)

At this stage the sun resembles a crescent moon, but the best thing to do is look at the surrounding countryside. Colors are saturated; shadows are stark; contrast is boosted; the shadows of trees and bushes contain innumerable strange crescent shapes. Ordinary objects such as trees and houses seem unfamiliar, as if illuminated by a star other than the sun. Everyday scenery has been transformed into something extraordinary.

Expectation fills the air. Then a minute or two before totality, shimmering dark lines suddenly wiggle over all white surfaces, such as sand or a sheet spread on the ground. These are called shadow bands, and they can't be photographed! If you try, your video or still images will show the white substance or object without any wavy bands at all. The rather anticlimactic reason for this is simply that shadow bands have extremely low contrast. Because they shimmer, the eye readily picks them out. But they lie below the contrast required to show up in a photographic image.

Then comes totality, which can last anywhere between one

second and around seven minutes. Now you take off your welding goggles and look at the sun directly. The bright stars come out. The sun's corona leaps across the sky, much farther than you expected. Its delicate wispy structure, following the sun's normally invisible magnetic-field lines, depends on the part of the solar cycle you're in. At a glance you'll know if you're at sunspot minimum or maximum: during the latter period the corona is round and symmetrical, as if the sun's springs have been wound up tightly and all the power held in place is ready to pop. But a quiet sun, paradoxically, lets go with long, irregular coronal streamers. Whenever it's seen, the glow is obviously that of a light different from anything nature normally offers. There is a logical reason for this, too: the sun's corona is by far the hottest thing the human eye can observe. It's made of plasma—broken fragments of atoms—rather than the whole atoms that comprise the solar surface and everything else around us on earth.

It's an experience that does not seem of this life or this world. "The home of my soul" is how one eclipse watcher described it to me. But why? What has really happened? It's obviously not simply a matter of the sun's visible light being blocked. Its invisible rays are extinguished, too. (As Victor Hess discovered during a 1912 near-total eclipse, when he went up in a balloon to measure the sun's radiation, cosmic rays do not decrease when the sun is blocked. But many other energies do indeed vanish.) Solar ultraviolet energy drops to zero. So does infrared radiation, whose absence starts to be felt long before totality arrives. With the drop in infrared energy, clouds, rocks, and the air just above the ground are suddenly cooled. This chill creates a pressure difference that manifests itself as a haunting eclipse wind. Moreover, the decreasing temperature as the sun is steadily blocked can shrink the gap between the temperature and the dew point, allowing clouds to suddenly form. That's what happened during the Siberian eclipse of the 1980s, with exasperating consequences, as the large international party of professional astronomers who had gathered to observe the event saw nothing when thick clouds materialized. They had meticulously planned for what the sun's visible rays would do—but they'd neglected its invisible rays!

When the eclipse is over, observers immediately start thinking about how they can get to the next one. If you missed being in the narrow ribbon of totality in the US in August 2017, don't fear; book a holiday to Chile or Argentina for the next solar eclipse on 2 July, 2019, where the moon is predicted to block out the sun for over four and a half minutes. And there's not long to wait between each eclipse if (or, more likely, when) you catch the bug: Antarctica in 2021; the US in 2024;* Spain and Egypt in 2027; and Australia in 2030. The UK, on the other hand, will have to wait until September 23, 2090, before seeing another one—if it's anything like the overcast 1999 Cornwall eclipse, however, it's more likely to be drizzle than drama.

In fact, wherever you go, be sure to factor in likely cloud cover. A good rule is that, in most places, mid-mornings tend to be clearer than mid-afternoons. I know someone who went to seven total eclipses but was clouded out of four of them. There are even several people who *chose* to view the 1999 eclipse in Cornwall, instead of in the crystal clear skies of Turkey. This is a case in which considerations of convenience—or, perhaps, having friends or relatives in a particular location—can steer us wrong.

* If you want to make specific plans, the track of the event can be found at the NASA eclipse website: https://eclipse.gsfc.nasa.gov/SEgoogle/SEgoogle2001/SE2024Apr08Tgoogle.html.

CHAPTER 25

ETs May Be Broadcasting, but What's Their Number?

Human beings belong to a lonely hearts club. We have no one to talk to except ourselves. We love our dogs and cats, but they're not conversationalists. The notion of advanced (and perhaps loquacious) extraterrestrials living on another planet carries an appeal that goes back centuries. Galileo was not alone in believing that the moon's dark blotches were oceans—and any place you have water, you can certainly have life.

The notion of smart creatures in other worlds grew so popular that in the early nineteenth century, William Herschel, our discoverer of infrared rays, believed that even the sun's surface was home to humanoid life. He wrote that the sun's outermost heat-emitting layers are blocked by thick, inner protective clouds, like a hazmat shield, allowing the presumed meadows on the sun's surface to remain pleasant picnic locales.

Near the end of that century, Guglielmo Marconi and Nikola Tesla both attempted radio communication with Mars. By then scientists realized that visible light was an inadequate medium for sending messages between worlds. Radio signals were much more logical. When Mars came unusually close to Earth, from

August 21 to 23, 1924, the notion of a National Radio Silence Day gained so much support that—at least in some areas—total radio silence was maintained for the first five minutes of every hour for thirty-six hours while a powerful dirigible-borne radio receiver from the United States Naval Observatory listened in vain for any signals from the Red Planet.

By the 1950s, intelligent ET life in our own solar system no longer seemed likely, and our attention eventually turned to the many planets believed to orbit some of the four hundred billion stars of our Milky Way galaxy. The issue, then as now, was: how many of these distant suns have planets—and how many of *those* can be expected to harbor intelligent life?

The answer to the first question only arrived at the beginning of this century. Two separate methods of hunting for exoplanets (planets that orbit a star outside our solar system) have already revealed thousands of them. It is now obvious that in our own galaxy alone there must be at least one billion earthlike worlds. That definition applies to planets of roughly our mass and temperature that orbit their stars at just the right distance to allow liquid water to exist on their surfaces.

But how many of those billion planets are home to intelligent life? Astronomers tend to be optimistic. After all, they've detected amino acids, the building blocks of life, in many nebulae. It's no longer much of a stretch to believe in panspermia—the theory that life's precursors (if not actual simple microscopic forms of life itself) may lie in protected cracks and crevices of meteoroids roaming the vast hallways of interstellar space and plant their seeds during collisions with planets. If this is so, life should be plentiful. Only a small minority of astronomers take the opposite viewpoint, which is that life's genesis is an extremely rare phenomenon and that essentially Earth is the site of a miracle.

If we'd like to get in touch with any of these purported aliens, there are two methods, which essentially boil down to speaking versus listening. The former strategy means sending signals into space announcing our existence and position, then waiting to see if anyone replies. This has already been done to a small degree. In 1974 a three-minute message was broadcast using the giant Arecibo radio telescope in Puerto Rico, aimed at the rich globular star cluster M13, announcing our position in space. Given the light-speed travel time of those radio waves, the message will arrive at the one million stars of that cluster twenty-five thousand years from now. If any aliens there respond promptly, we'd expect to get their "Hi! Yes, we're good. How are you?" sometime around the year 52000.

In some circles, the idea of sending active messages into the cosmos has generated alarm. For example, the noted physicist Stephen Hawking warns that in our own earthly experience, whenever a civilization has let itself be discovered by a more technologically advanced culture, the results have rarely been favorable for the former.

Just as it's not a good idea to go around shouting in the jungle if you don't know what's around you, perhaps we should not assume that all technologically advanced extraterrestrials would have a benign attitude toward us. Says Hawking, "It would be wiser for us to lay low." Not surprisingly, many scientists regard this stance as unduly paranoid.

A less controversial way to search for extraterrestrials is to use our radio antennas to listen for either deliberate or accidental transmissions. After all, we earthlings have been broadcasting radio and television signals since the 1920s. It's nearly a cliché to say that the *I Love Lucy* TV show is still zooming outward and has already reached a distance of sixty light-years from our planet. Its signal has reached hundreds of nearby stars and presumably a similar number of exoplanets. Not popularly appreciated, however, is that these signals are very weak. Aliens equipped with our own current best radio telescope technology would be unable to detect our presence from these commercial transmissions, even if they lived on a planet orbiting the very nearest star. But what if they were far more advanced than we are? What if they could both listen and broadcast at an extraordinarily sensitive level?

Maybe alien signals are mere incidental transmissions just like ours, since we certainly did not broadcast *Hawaii Five-0* and *The Beverly Hillbillies* so that they could be enjoyed in other worlds. Moreover, our radio and television towers transmit sideways, with the goal of reaching audiences stretching horizontally into the distance. Our commercial broadcasts are never aimed upward through the thinnest part of the atmosphere, toward space. Nor are they intended to be widely dispersed; they're preferentially beamed toward population centers. Nor do they start off with interplanetary-level power.

But the situation would be different if aliens were deliberately sending signals in the hope of being discovered. Starting in the 1960s, many listening projects have been funded with the goal of detecting exactly such alien transmissions.

The first question is: what should we listen for? Of course we're looking to detect invisible light. But at what wavelength and frequency? Essentially, what's their phone number?

The Crab Nebula sends out energy from almost the entire electromagnetic spectrum. If extraterrestrials are deliberately transmitting "hello" signals, which frequency would they choose for their broadcasts? *(NASA)*

Realistically, we cannot listen to frequencies lower than one gigahertz, meaning the low microwave and long radio-wave part of the spectrum. Any lower than that, and there's simply too much noisy competition and interference with our own earthly transmissions. That limits us—a position analogous to only looking for lost keys under a streetlight, as one noted ET investigator put it to me. But even excluding frequencies lower than one gigahertz, that still leaves us the vast majority of the electromagnetic spectrum.

We assume that extraterrestrials would send radio waves, especially because (unlike visible light) they can be easily broadcast in all directions at once instead of focused in a single direction. But ETs might beam in the infrared section of the spectrum because such transmissions could be more concentrated. Indeed, some people have even suggested that the supermysterious gamma-ray bursts that pop out in the sky around once a day and usually last for just a few seconds might be concentrated high-power streams of artificially broadcast information from truly distant aliens in other galaxies.

In 2016, I spoke with Seth Shostak, the famed director of the SETI Institute, which has been conducting its search for extraterrestrial intelligence for decades.

"We get about four hundred letters a day," he said with a sigh, "mostly from people with ideas about what frequencies we should be listening to. For example, some think that any alien intelligence would deliberately choose the numbers in pi, knowing that all smart life forms would recognize it as the universal figure that expresses the relationship between every circle's circumference to its diameter. In other words, we should tweak our radio telescopes to try to detect 3.141592 gigahertz."

Others suggest some multiple of pi or things like pi times the 1,420-mHz frequency signature of hydrogen. In other words, asks Shostak rhetorically, "What would ET use as a hailing channel?" The problem with the electromagnetic spectrum is that the number of frequencies of invisible light are essentially limitless, while we keep listening to one or another very focused band.

SETI's ongoing efforts are the most famous, but they do not

represent the first time we've hunted for invisible rays from extraterrestrials. In 1960, astronomer Frank Drake, who would go on to become a professor at Cornell University, used the eighty-five-foot radio telescope at the National Radio Astronomy Observatory in Green Bank, West Virginia, to try to detect messages from any unknown planets orbiting the nearby stars Epsilon Eridani and Tau Ceti, which are solar analogues and therefore not too fiercely hot. Called Project Ozma, the program was named after Princess Ozma, ruler of the fictional land of Oz, which helped the project garner much mainstream media attention. Drake "listened" at the 1.420 gigahertz frequency, where hydrogen emits its famous twenty-one-centimeter-wavelength emission hiss—a logical place on the spectrum and a place that erudite aliens might know would be the natural focus for intelligent life. But his efforts produced no results.

In 1971, NASA authorized funding for a SETI study, and in the 1980s a project named Sentinel used Harvard's eighty-five-foot-diameter radio telescope hooked up to a spectrum analyzer that permitted listening to 131,000 frequencies at the same time.

By the mid-1990s, a new search, called BETA (Billion-Channel Extraterrestrial Assay), had the ability to receive 250 million channels simultaneously. Along with follow-up projects, the search continued until a near-hurricane-level storm blew over and destroyed the radio telescope in 1999.

Further listening for invisible light signals has made use of ever-improving technology, including spectrum analyzers capable of detecting fifteen million different frequencies simultaneously. These projects, some privately funded, some administered by universities or NASA, with names like MOP (Microwave Observing Project), SERENDIP (Search for Extraterrestrial

The Allen Telescope Array, which the SETI Institute uses in its hunt for extraterrestrial life. *(Seth Shostak)*

Radio Emissions from Nearby Developed Intelligent Populations), Project Phoenix, and Breakthrough Listen, are ongoing. And although hundreds of millions of potential technological (i.e., nonnatural) signals have been detected, all have thus far been attributed to noise and earthly satellites, or they have vanished too quickly to be properly analyzed.

In addition to the idea that an extraterrestrial intelligence might be sending us deliberate signals is the notion of "incidental" signatures from advanced technologies or even from something as simple as the infrared heat from a planet's city lights and engine exhausts pulsing on and off as that world rotates.

As of 2016, the number of stars we've looked at carefully, meaning those we've monitored thoroughly on lots of frequencies, is in the low thousands. So far, like Pavlov's dogs, we've been patiently salivating but haven't yet received a single biscuit.

It's possible that we're wildly wrong about the likelihood of intelligent life out there. In the 1950s, the famed Italian physicist Enrico Fermi told his colleagues that if advanced civilizations are common in our galaxy, they should be detectable. But if that's true, he asked, "Where is everybody?" This Great Silence, as some are now calling it, may suggest that advanced extraterrestrial life may be much rarer than we supposed. Such pessimism isn't far-fetched. Life originating randomly from chemical combinations requires astoundingly rare sequences of events, which is why Fred Hoyle, who coined the term *big bang,* characterized the accidental genesis of life as akin to a tornado sweeping through a junkyard and assembling a working jumbo jet.

Another explanation might be that others' intelligence is not at all like ours. Just because we like to communicate via microwaves and radio waves doesn't mean that others do. And just because we're drawn to the notion of strapping ourselves inside a rocket and blasting into the hostile emptiness of space doesn't mean that other intelligences find such activity in any way appealing. Perhaps they have sufficient peace of mind to stay put.

What if ETs can communicate perfectly well among themselves without using waves of energy? Indeed, right here on earth, the most intelligent beings besides ourselves—dolphins—show not the slightest desire to hurl themselves outside the seas and beyond the planet. Nor do they show a desire to build electronic devices. What if they're closer to the norm for ET intelligence than we are? What if all our listening to millions of frequencies of invisible light is a futility born of our own anthropocentrism?

There's no way to know. But we're not going to stop.

We'll keep trying new phone numbers until someone picks up the receiver.

CHAPTER 26

Does Light Have a Bright Future?

Two hundred years after William Herschel stumbled upon the invisible spectrum, there's no sign that the flood of invisible light will ever grow less intense. On the contrary, even more unseen energies will be pumped through our homes in the coming years. Might these energies displace the visible wavelengths? Might society heed environmentalists' calls to cut down on "waste lighting" and light pollution to preserve the glories of the natural night sky, all while ramping up the intensity of microwaves, radio waves, and other invisible electromagnetic frequencies?

Already plans are afoot to use solar-powered drones to beam continuous widespread Wi-Fi signals over large areas so that "hot spots" are no longer spots but rather the norm everywhere. In short, you'd be able to use your smartphones and their apps even if you were hiking in a national park.

Meantime, as SiriusXM satellite radio has demonstrated, satellites can and do beam microwaves and radio waves in blanket coverage that is blocked only by steep canyons and rock formations. The overall effect on our health? Probably somewhere between zero and negligible, although some say that "probably" isn't enough assurance if you're exposing absolutely everyone, including young children, to body-penetrating energy waves.

Meanwhile, the use of hard, ionizing sections of the electromagnetic spectrum—we're talking X-rays—continues to increase with the still-growing global popularity of CT scans. Articles in medical journals published since 2014 estimate that as many as 2 percent of all cancers are caused by medical X-rays, mostly from CT scans. Put another way, a single whole-body CT scan probably increases your cancer-death risk by one chance in two thousand. Happily, the awareness of this huge modern source of ionizing radiation is making many physicians step back and question whether CT scans are really necessary or whether simple X-rays may suffice in many situations. The awareness of the problem has arrived: we are now in stage 2—which is acting on it.

Cell phones are another story. As their use goes from commonplace to ubiquitous, we know that virtually everyone is now exposed to microwaves. Those who hold their phones flush against their heads get the highest exposure by far. Those who simply text, or use headphones or earbuds, receive very little of this radiation. Fortunately, the phones themselves have been reducing their broadcast-output signal strength. As of 2016, the amount of microwaves in the air from cell phones is only around half what it was just fourteen years earlier. Ongoing studies examine animal exposure over long durations. As noted earlier, studies on humans have been mostly reassuring, but until certainty is reached, it might be wise to use the speakerphone feature.

When we speak of cosmic rays, we're talking about invisible particles rather than rays, but nonetheless it is interesting that the strength and frequency of the sun's activity—its flares, coronal mass ejections, and sunspots, all of which emit a steady "wind" of charged particles—have been declining since the late 1990s. The sun's current cycle, number 24, is the wimpiest in our lifetimes, and most solar researchers believe that the sun has entered an extended period of inactivity.

At first this may seem reassuring to those who are wary of cosmic rays and their by-products, including potentially cell-damaging muons, two hundred of which penetrate our bodies every second. A quiet sun should mean fewer cosmic rays. However, it turns out that the real consequence is just the reverse.

That's because the most intense cosmic rays are not of solar origin but rather come from beyond the solar system; these are primarily the rays that affect animal life on earth. And most of these are deflected at the heliopause—the outer edge of the solar system—*but mainly when the sun is active.* In short, the sun's current quietude has reduced the effectiveness of its barrier against interstellar invaders. The situation is allowing far more of those powerful cosmic rays to penetrate all the way to our planet's atmosphere—and even to the ground. This is experienced most by those who live at high elevations, as in Colorado, and by those whose homes are remote from the equator, as in Alaska.

As for ultraviolet radiation, the sun's quietness has also reduced UV intensity. Although each sunspot minimum (a quiet period in solar activity, which is predicted to occur again in 2022) reduces visible solar insolation—the brightness of the sun's energy as it strikes our world—by only a single watt per square meter of the earth's surface (compared to the sun's brightness at its maximum), it also cuts the sun's UV emissions by more than 10 percent.

It is also true that the ozone hole at extreme southern latitudes continues to let more UV radiation bleed through than will be the case if and when that region of ozone depletion is patched. Still, as we've seen, UV radiation is less of a problem than was feared a few decades ago, at least in terms of melanoma genesis. The advice stands: don't hide from the sun. Instead get as much sunlight as you can *without burning.*

As for gamma rays, they will continue to be used for food irradiation—in the best-case scenario, irradiation will be limited to a small minority of foodstuffs, such as spices. Besides people who work at nuclear power plants and people with advanced cancer (for whom radiation is used as a palliative), no humans or animals are exposed to gamma rays.

Infrared radiation continues to be used for garage-door openers, in heat lamps, in fire-detection equipment, in spy cameras, and in other products and gadgets. No harm to humans has ever been associated with this part of the spectrum. Possibly the most promising new IR technology involves restoring sight to the blind. One such product went on the market in 2016, and the company claims it will give totally blind people vision of at least 20/250, which would be sufficient to read the topmost line—that enormous *E*—on a standard Snellen chart.

How does it work? A surgeon first implants a small silicon chip containing 150 electrodes on the retina. When the blind person puts on the system's dark glasses, an integrated video camera sends images to a portable computer. A connected "pocket processor" converts that recording into an infrared image, which the glasses then beam into the eye. Pulses activate electrodes in the implant, and the optic nerve carries images to the brain. This wondrous concept has, as of this writing, been successful in its first medical applications.

In the spring of 2016 came the surprising announcement from Europe that even a basic property of light can have new, unexpected aspects. Specifically, light has an angular momentum, or rotational inertia, which science has exploited in various ways, including for the storage of quantum information. Viewed in the normal three-dimensional way, a photon's angular momentum has a value that is expressed as a whole number, or integer, and is never fractional. But researchers have shown that when considered in some reduced dimensions, a photon's momentum can have a fractional value, and no doubt this new finding will be exploited in future technological applications, such as coding methods that protect sensitive stored information. The point is, when the subject is light—seemingly so basic to reality—we are *still* not close to full understanding. Despite the quantum leaps of knowledge we've made since the days of Hendrik Lorentz, at the dawn of the twentieth century, light always seems to have new surprises for us.

As for spooky energies—dark energy and vacuum energy and zero-point energy—these large-scale facets of nature may also be fully present in our homes. As far as we know, their flux—their environmental intensity—doesn't vary over the course of human lifetimes. Their effects on us are unknown but likely are minimal. Being ubiquitous, they are analogous to the oxygen and water vapor surrounding us. Logic dictates that anything that is everywhere in equal measure cannot be harmful to life forms occupying a finite position.

Even as our knowledge of these newly found ultrapowerful unseen energies increases, we don't know whether they will be exploited in our lifetimes. At the moment we don't know how we'd even take our first steps toward harnessing them. Still, the lure of an absolutely unlimited energy source keeps some of our greatest minds up at night. Throughout history, science has taught us to "never say never." These energies could be invisible El Dorados of beneficence—fountains of limitless power and deliverers of vastly boosted living standards.

After all, the invisible light that *has* been discovered, studied, and harnessed has profoundly changed the way we live. We rely on the invisible parts of the spectrum for communication and food preparation. We rely on them to see the insides of our own bodies and to study the far reaches of outer space. Yes, the growing use of drones and Wi-Fi and other technologies has ramped up the flux of microwaves and radio waves in our homes. But the main takeaway from this book is certainly not that we should be fearful. Rather, my purpose has simply been to open a window onto the enormous universe of omnipresent energies, most of them benign, that fill every moment of our lives.

I hope the tales of their discoveries, the ways in which they changed civilization, and the ways in which they affect us today will do more than inspire us to salute the men and women who dedicated their lives to advancing this science and these technologies. Given that a world of unseen power and potential surrounds us—of which we have only scratched the surface—I also hope we will add a good measure of wisdom to our cleverness as we explore the unseen lights that blaze brilliantly in realms beyond what our senses can ever perceive.

Index

About the Author

BOB BERMAN, one of America's top astronomy writers, contributed the popular "Night Watchman" column to *Discover* for seventeen years. He is currently a columnist for *Astronomy,* a host on the weekly *Strange Universe* show on WAMC Northeast Public Radio, and the astronomy editor of *The Old Farmer's Almanac*. His books include *Zoom* and *The Sun's Heartbeat*. He lives in Willow, New York.

SkyManBob.com